Łukasiewicz Logics and Prime Numbers

ŁUKASIEWICZ LOGICS

AND

PRIME NUMBERS

Alexander S. Karpenko

*Department of Logic, Institute of Philosophy,
Russian Academy of Sciences, Moscow*

Luniver Press

Published by

Luniver Press
Beckington, Frome BA11 6TT, UK

www.luniver.com

Alexander S. Karpenko
Łukasiewicz Logics and Prime Numbers
(Luniver Press, Beckington, 2006).

First Edition 2006

British Library Cataloguing in Publication Data
A catalog record is available from the British Library.

Library of Congress Cataloguing in Publication Data
A catalog record is available from the Library of Congress.

ISBN-13:   978-0-9551170-3-9
ISBN-10:   0-9551170-3-8

Dedicated to my wife Tatyana Gobzeva,
without whom there would not have long been
not only me, but also logic with prime numbers.

Alexander S. Karpenko, D.Sci. (1991), PhD (1979), is Head of Department of Logic, Institute of Philosophy, Russian Academy of Sciences, and Professor of Moscow State University.

CONTENTS

Introduction

*"...the philosophical significance of the [many-valued]
systems of logic treated here might be at least as great as
the significance on non-Euclidean systems of geometry."*

J. Łukasiewicz, 1930.

*"...there is no apparent reason why one number is prime
and another not. To the contrary, upon looking at these
numbers one has the feeling of being in the presence of
one of the inexplicable secrets of creation."*

D. Zagier, 1977.

The present book is a substantially rewritten English version of its Russian predecessor ([Karpenko, 2000]).[1]

The title of the book may appear somewhat strange since, at first glance, what can logic and prime numbers have in common? Nevertheless, for a certain class of finite-valued logics such commonalties do exist – and this fact has a number of significant repercussions. Is there, however, any link between the doctrine of logical fatalism and prime numbers?

Jan Łukasiewicz (1878-1956) was a prominent representative of the Lvov-Warsaw School (see [Woleński, 1989] for details) and the originator of mathematical investigation of logic within that school. His criticism of Aristotle's fatalistic argument laid the ground for the historically first non-classical, three-valued, logic. Its properties proved to be – for Łukasiewicz's time – somewhat shocking; its subsequent generalizations for an arbitrary finite and – further still – the infinite cases showed that the modeling of the infinite and the finite on the basis of Łukasiewicz many-valued logics yields results that justify the claim that, by the end of the twentieth century, there have taken shape and are now rapidly growing two distinct and significant trends in the contemporary symbolic logic: Łukasiewicz infinite-valued logic $Ł_\infty$ (see [Cignoli, D'Ottaviano and Mundici, 2000]) and Łukasiewicz finite-valued logics $Ł_{n+1}$ – the logics discussed in this book. While in the former case the beauty of the subject arises out of consideration of different (but equivalent) algebraic structures serving as counterparts of the logic as well as out of its various applications; in the latter, we enter the mystical

---

[1] See A. Adamatzky's English extensive review of this book [Adamatzky, 2004].

4

world of prime numbers, the world that proves to be connected with the functional properties of $Ł_{n+1}$.

The book consists of three parts, dealing with, respectively, (1) Łukasiewicz finite-valued logics $Ł_{n+1}$; (2) their link with prime numbers; and, lastly, (3) the numeric tables illustrating the link described in part (2).

Chapter I is an elementary introduction to the two-valued classical propositional logic C_2. It is worth noticing that Łukasiewicz two-valued logic $Ł_2$ is nothing else than C_2. It means that all Łukasiewicz many-valued logics are generalizations of C_2. Chapter II describes the origin and development of Łukasiewicz three-valued logic $Ł_3$ and indicates the connection between $Ł_3$ and the problem of logical fatalism. Some surprising and unexpected properties – such as the failure of "the laws" of excluded middle and non-contradiction – of $Ł_3$ are also considered there; that consideration makes apparent that, as soon as we introduce some novelties into the classical logic, there arises a thorny problem of what interpretations of the logical connectives and of the truth-values themselves are intuitively acceptable. (This problem, in turn, leads to the problem of what is a logical system – all the more so, given the at first glance surprising fact that $Ł_3$, as well as any other $Ł_{n+1}$, can be axiomatically presented as a restriction of a Hilbert-style axiomatization of C_2 and also as an extension of a Hilbert-style axiomatization of C_2.)

In Chapter III we consider some properties of $Ł_{n+1}$, including degrees of cardinal completeness of $Ł_{n+1}$'s (first studied by A. Tarski in 1930) – the property that allowed us the first glimpse of a connection between $Ł_{n+1}$'s and prime numbers. Towards the end of Chapter III, we propose an interpretation of $Ł_{n+1}$ through Boolean algebras.

In our view, neither the axiomatic nor the algebraic (nor, for that matter, any other semantic) approach can bring out the uniqueness and peculiarity of Łukasiewicz finite-valued logics $Ł_{n+1}$. All these approaches we call external, as opposed to the approach considering $Ł_{n+1}$'s as functional systems. We believe that only the latter approach can help us decipher the essence of $Ł_{n+1}$'s. It was exactly this approach that allowed to discover that functional properties of $Ł_{n+1}$ are highly unusual. V.K. Finn was the first to note this in his brief paper "On classes of functions that corresponded to the n-valued logics of J. Łukasiewicz" ([Finn, 1970]). One repercussion of Finn's work is that the set of functions of the logic $Ł_{n+1}$ is functionally precomplete if and only if n is a prime number. Finn's result is discussed in Chapter IV, which is *crucial for our consideration* because it provides a bridge between the first and the second parts of the book. (It should be noted that Finn's result – which was later independently re-discovered – is both the foundation of and the primary inspiration for the writing of the present book.)

The Finn's led to an algorithm mapping an arbitrary natural number to a prime number with the help of the Euler's totient function $\varphi(n)$, thus inducing a partition of the set of natural numbers into classes of equivalence; each of thus obtained classes can be represented by a rooted tree of natural numbers with a prime root. That algorithm, in turn, led to an algorithm based on some properties of the inverse Euler's totient function $\varphi^{-1}(m)$ mapping an arbitrary prime number p to an equivalence class equivalence of natural numbers. Chapter V contains thus obtained graphs for the first 25 prime numbers as well as the canceled rooted trees for prime numbers from 101 (No. 26) to 541 (No. 100). Thus, each prime number is given a structure, which proves to be an algebraic structure of p-Abelian groups.

Some further investigations led to the construction of the finite-valued logics $\mathbf{K_{n+1}}$ that have tautologies if and only if n is a prime number ($\mathbf{K_{n+1}}$ are described in Chapter VI). The above statement can be viewed as a purely logical definition of prime numbers. $\mathbf{K_{n+1}}$ happen to have the same functional properties as $\mathbf{L_{n+1}}$ whenever n is a prime number. This provided the basis for constructing the Sheffer stroke operator for prime numbers. (In this construction, we use formulas with 648 042 744 959 occurrences of the Sheffer stroke.) It is interesting that a combination of logics for prime numbers helps discover *a law of generation of classes of prime numbers*. As a result, we get a partition of the set of prime numbers into equivalence classes that are induced by algebraic-logical properties of Łukasiewicz implication; all prime numbers can be generated in such a way.

Finally, in Chapter VII we give what we consider to be an ultimate answer to the question of what is a Łukasiewicz many-valued logic. Its nature is purely number-theoretical; this is why it proves possible to characterize, in terms of Łukasiewicz logical matrices, such subsets of the set of natural numbers as prime numbers, powers of primes, odd numbers, and – what proved to be the most difficult task – even numbers. (In that last case, we also try to establishing a link with Goldbach's conjecture concerning the representation of every even number by the sum of two prime numbers.)

The third part of the book is made up of the numerical tables never previously published. Table 1 contains the values of the cardinal degrees of completeness for n-valued Łukasiewicz logics ($n \leq 1000$). (Some natural numbers, in this respect, happen to form a special "elite".) In table 2 the cardinality values of rooted trees and of canceled rooted trees are given for p

≤ 1000. Table 3 in this book gives the values of function $i(p)$, for $p \leq 1000$, which partitions prime numbers into equivalence classes.

In algebraic terms, in this book we investigate the clones of finite MV-chains.

Concluding remarks contain some metaphysical reflections on extensions of *pure logic*, on prime numbers, and on fatalism and continuality, as well as on the possible connections of these themes with Łukasiewicz logics.

Acknowledgements

The construction of the tables making up the third part of this book would be impossible without the computer programs written especially for this book by my friend and colleague Vladimir Shalack. I would like to warmly thank Vladimir, without whose contribution this book would be incomplete. No less warm thanks are due to Dmitry Shkatov, without whose editing of the author's English the book in its present form would be impossible.

Alexander S. Karpenko,
Moscow, 2005

I.   Two-Valued Classical Propositional Logic

I.1.   Logical connectives. Truth-tables

Propositional logic is the part of modern symbolic logic studying how complex propositions are formed out of simple ones and their interrelations. By contrast with predicate logic, simple propositions are considered in propositional logic as atoms, that is their inner structure is not taken into account and we only pay attention to how simple propositions are combined into complex ones with the help of various conjunctions. By proposition is meant "a written or uttered sentence which is declarative and which we agree to view as being either true or false, but not both" (see chapter II, "Classical Propositional Logic," of the excellent book by R.L. Epstein [Epstein, 1980, p. 3]).

There are many ways to join propositions to form a new, more complex, proposition in natural language. We single out five well-known conjunctions: 'not', 'if … then…', 'or', 'and', 'if and only if …'. The process of symbolization (formalization) of natural language by means of propositional logic consists in the following. Atomic (simple) propositions are replaced by *propositional variables p, q, r, …*, possibly with indices; the five above-mentioned conjunctions are represented as *logical (or, propositional) connectives*: $\neg$ (negation), $\wedge$ (conjunction)[2], $\vee$ (disjunction), $\supset$ (implication), and $\equiv$ (equivalence), respectively. Lastly, we use the parentheses '(' and ')' to facilitate grouping propositions in various ways. Variables A, B, C, … possibly with indices range over all the propositional variables and complex expressions formed thereof; these variables are also referred to as *metavariables*. Analogous to sentences of English in the language of propositional logic are *well-formed formulas (wffs)*. There follows the definition of wffs:

(1) Every propositional variable is a wff.

(2) If A is a wff, then $(\neg A)$ is a wff.

(3) If A and B are wffs, $(A \supset B)$, $(A \vee B)$, $(A \wedge B)$, and $(A \equiv B)$ are wffs.

---

[2] Some authors use & rather than $\wedge$ to denote conjunction.

9

(4) No other string of symbols is a wff.

To simplify our notation, we will leave out the outermost parentheses of a formula. We denote the set of all wffs by *For*. Henceforth, by formulas we always mean wffs.

The classical logic is based on the following two major assumptions:

(I)   *The principle of bivalence.* Each atomic proposition is either true or false and not both. "Truth" and "Falsehood" are called *truth-values* and are denoted respectively as 'T' and 'F', or '1' and '0'.

(II)   *The principle of extensionality.* The truth-value of a complex proposition depends only on its connectives and the truth-values of its component atomic propositions. Thus, the propositional connectives stand for *truth-functions.*

The following question arises: what truth-functions correspond to the propositional connectives?

A convenient way to represent truth-functions is through tables with all possible combinations of values of arguments (propositional variables) on the left and the values of the function on the right, as in the following tables:

p	$\neg p$
1	0
0	1

p q	$p \supset q$	$p \vee q$	$p \wedge q$	$p \equiv q$
1 1	1	1	1	1
1 0	0	1	0	0
0 1	1	1	0	0
0 0	1	0	0	1

Thus, for example, $p \supset q$ is 0 if and only if (iff, for short) p is 1 and q is 0; otherwise $p \supset q$ is 1. Such tables are usually called *truth-tables*, and the propositional connectives definable by means of such tables are said to be *truth-functional*; the connectives definable by truth-tables with only two possible arguments, say 1 and 0, are said to *classical* (thus, all the above truth-tables represent classical connectives).

It is easy to find out how many different classical connectives of m arguments there are: the number of rows in a table for a truth-function of m arguments is 2^m, and there are two possible values of the function for each row can: 1 and 0. Hence, the number of classical connectives of m arguments is 2^{2^m}. Thus, there are 4 unary classical connectives, 16 binary classical connectives, etc.

I.2.   The laws of the classical propositional logic

Every well-formed classical formula corresponds to a truth function, which can be graphically represented by a truth-table. In other words, every classical formula can be seen as a function of variables ranging over the set $\{0, 1\}$, and hence the truth table method can be extended to all formulas in *For*. A function $v : For \rightarrow \{0, 1\}$ is a *logical valuation* of the set of formulas *For* if, for any $A, B \in For$,

$$v\,(\neg A) = 1 \quad \text{iff } v(A) = 0$$

$$v\,(A \supset B) = 0 \quad \text{iff } v(A) = 1 \text{ and } v(B) = 0$$

$$v\,(A \vee B) = 0 \quad \text{iff } v(A) = v(B) = 0$$

$$v\,(A \wedge B) = 1 \quad \text{iff } v(A) = v(B) = 1$$

$$v\,(A \equiv B) = 1 \quad \text{iff } v(A) = v(B).$$

Among the formulas of *For*, we can distinguish the ones that have 1 as the value of every row of their truth-tables. Such formulas are called *tautologies*, and 1 is called a *designated* truth-value. Thus, every valuation assigns to a tautology a designated truth-value.

Tautologies play a paramount role in propositional logic. They are "the laws of logic," the formulas that are true in virtue of their symbolic form alone; in other words, the truth value of a tautology does not depend on the values of its atomic propositions. It is easy to check that the following formulas are tautologies:

(1) $p \supset p$,

(2) $p \vee \neg p$,

(3) $\neg(p \wedge \neg p)$.

As I.M. Copi and C. Cohen wrote in the section titled *The Three "Laws of Thought"* of their popular undergraduate-level logic textbook [Copi and Cohen, 1998, p. 389], "Those who defined logic as the science of the laws of thought have often gone on to assert that there are exactly three fundamental or basic laws of thought necessary and sufficient for thinking to follow if it is to be 'correct'." These laws have traditionally been referred to as "the law of *identity*," the law of *excluded middle*" (*tertium non datur*), and "the law of *contradiction*" (sometimes "the law of non-contradiction"). They had already been formulated by Aristotle in an informal way. There are some alternative formulations of these laws. Let's take a look at such alternatives for the second and the third of the above laws:

The law of excluded middle can be taken as asserting that *two contradictory propositions are not false together; that is, that one of them must be true.*

The law of contradiction can be taken as asserting that *two contradictory propositions are not true together; that is, that one of them must be false.*

Aristotle in his *Metaphysics* stressed that the law of contradiction is "the most certain of all principles." However, as we will see in the next chapter, both of these laws are jettisoned in Łukasiewicz three-valued logic.

Let's list some more of classical tautologies.

(4) $\neg\neg p \equiv p$ (*double negation*)

(5) $(p \supset q) \supset (\neg q \supset \neg p)$ (*contraposition*)

(6) $(\neg p \supset \neg q) \supset (q \supset p)$ (*inverse contraposition*).

In what follows, we will also need some purely implicational tautologies (of particular importance to us will be the following implicational version of contraction law):

(K) $p \supset (q \supset p)$ (*weakening*)

(S) $(p \supset (q \supset r)) \supset ((p \supset q) \supset (p \supset r))$ (*self-distribution*)

(B') $(p \supset q) \supset ((q \supset r) \supset (p \supset r))$ (*transitivity*)

(C) $(p \supset (q \supset r)) \supset (q \supset (p \supset r))$ (*permutation*)

(W) $(p \supset (p \supset q)) \supset (p \supset q)$ (*contraction*).

The Classical Propositional Logic, C_2, is defined as the class of all classical tautologies.

Let's note that, using truth-tables, we can effectively calculate the truth-value of any propositional formula under any distribution of the truth-values of its constituent atomic propositions. Thus, we can find out whether an arbitrary formula is a C_2-tautology. Thus, we have a *decision procedure* for C_2.

Let's wrap this section up by citing some basic principles concerning the interrelations among classical tautologies (these principles are so basic that they are often referred to as the *rules of inference* of the classical propositional logic).

    1. *Modus ponens*. If A and $A \supset B$ are tautologies, then B is also a tautology.

    2. *Substitution*: If $A(p)$ is a tautology, then $A(B)$ is tautology, where B is any formula which is substituted uniformly for p (i.e., B replaces every occurrences of p in A).

I.3.   Functional completeness

We say that formulas A and B are *(logically) equivalent* if $A \equiv B$ is a tautology. Obviously, if A and B are equivalent, they have the same truth-tables.

    A system $\mathfrak{R}$ of propositional connectives of the classical propositional logic $\mathbf{C_2}$ is called *truth-functionally complete* if every connective representing a function on the set of truth-values $\{0,1\}$ can be defined through the connectives of $\mathfrak{R}$ alone. It can be shown that every propositional connective of $\mathbf{C_2}$ is definable through $\neg$, $\wedge$, and $\vee$, i.e. that the system of connectives $\{\neg, \wedge, \vee\}$ is truth-functionally complete. More precisely, for every connective $*$ of $\mathbf{C_2}$, we can construct, using only connectives $\neg$, $\wedge$, $\vee$, such a formula D that truth-tables for $*$ and D are identical.

Theorem 1. *The set of classical connectives $\{\neg, \wedge, \vee\}$ is truth-functionally complete* [Post, 1921].

    There follow some equivalences showing how the familiar classical connectives can be expressed in terms of each other:

$$p \vee q \equiv \neg p \supset q, \quad p \vee q \equiv (p \supset q) \supset q, \quad p \vee q \equiv \neg(\neg p \wedge \neg q);$$

$$p \wedge q \equiv \neg(p \supset \neg q), \quad p \wedge q \equiv \neg(\neg p \vee \neg q);$$

$$p \supset q \equiv \neg p \vee q, \quad p \supset q \equiv \neg(p \wedge \neg q);$$

$$(p \equiv q) \equiv (p \supset q) \wedge (q \supset p).$$

    It, thus, follows that the systems of connectives $\{\neg, \supset\}$, $\{\neg, \vee\}$, and $\{\neg, \wedge\}$ are also truth-functionally complete. It means that we can take any of them as the truth-functional basis for the classical propositional logic.

I.3.1.   Sheffer stroke

    Among all the connectives of the classical propositional logic, there are two forming a complete truth-functional system each on its own. One of these is the *Sheffer stroke* (discovered in 1913), written as $p|q$, which takes on value 1 iff both p and q are 0; that is, $p|q \equiv \neg p \wedge \neg q$; the following equivalences show that $p|q$ alone is sufficient as the truth-functional basis of the classical propositional logic: $\neg p \equiv p|p$, $p \wedge q \equiv (p|p) \mid (q|q)$. The other is *Pearce hand*, written as $p \downarrow q$, which takes on the value 1 iff either p or q is 0; that is $p \downarrow q \equiv \neg(p \wedge q)$; the following equivalences show that $p \downarrow q$ alone is sufficient as the truth-functional basis of the classical propositional logic: $\neg p \equiv p \downarrow p$, $p \wedge q \equiv (\neg p \downarrow \neg q)$.

Thus, to show that a logical connective * is truth-functionally complete, all we have to do is (*i*) define * through the initial connectives, and (*ii*) define the initial connectives through *. Some analogues of the Sheffer stroke of the classical logic we will consider later on in this book.

I.4.   Axiomatization. Adequacy

Alongside the concept of tautology, another concept crucial for logic is that of *logical consequence*. One of the basic tasks of logic is to show *what follows from what*. We say that "*B* logically follows from *A*, or *B* is a logical consequence of *A*" (symbolically, $A \vDash B$) if, in compatible truth-tables for *A* and *B*, formula *B* takes on the value 1 in every row where *A* is 1. It then follows that $A \vDash B$ iff $A \supset B$ is a tautology. If *A* is a tautology, we write $\vDash A$.

The above definition of logical consequence can easily be extended to collections of formulas; we write $\Gamma \vDash B$ to denote that formula *B* is a consequence of a set of formulas Γ. An example of the relation of logical consequence holding between a set of formulas and a single formula is the above-cited rule of modus ponens.

If have the concepts of tautology and of logical consequence defined (as we have done above for the classical propositional logic), then we have a *semantic presentation* of a propositional logic; then, a propositional logic can be simply identified with the set of its tautologies or with its relation of logical consequence. However, the question arises of how we can survey all of the infinite number of tautologies? To give an at least partial answer to this question, we should consider a *syntactic presentation* of propositional logic.

Under a syntactic approach, the formal language of propositional logic and the notion of wffs remain the same, but from the whole set of tautologies we single out a finite subset, elements of which we stipulate as *axioms*.

From the above considerations on functional completeness it follows that we can develop the classical propositional logic $\mathbf{C_2}$ on the basis of the system of connectives $\{\neg, \supset\}$. This is how $\mathbf{C_2}$ was initially set out by

G. Frege in 1879. Łukasiewicz in 1930 significantly simplified the axiomatization of C_2 suggested by Frege (see [Łukasiewicz and Tarski, 1970], p. 136):

Ax.1 $p \supset (q \supset p)$

Ax.2 $(p \supset (q \supset r)) \supset ((p \supset q) \supset (p \supset r))$

Ax.3 $(\neg p \supset \neg q) \supset (q \supset p)$.

Transitions from formulas to formulas are made according to the following rules:

R1. *Modus ponens*: $$\frac{A, A \supset B}{B.}$$

R2. *Substitution*: $$\frac{\vdash A(p)}{\vdash A(B).}$$

'$\vdash A$' means that A is a *theorem* of the system, i.e. that there is a proof of A from the axioms, i. e. there is a sequence $A_1, \dots A_m$ of formulas such that A_m is A and each A_i is either an axiom or is obtained from some of the preceding A_j's by one of the above inference rules. A detailed analysis of the above axiomatization can be found in [Church, 1956, ch. II].

Łukasiewicz suggested a very convenient notation for writing down the proofs of formulas [Łukasiewicz (1929), 1963]. Every thesis proved will be numbered and preceded by a proof line, which consists of two parts separated by an asterisk. For instance, let us consider the following proof.

Proposition 1. W, K $\vdash p \supset p$.

   1. **W.**

   2. **K.**

     1 $q/p * 2\ q/p - 3$,

   3. $p \supset p$.

Here, the first part of the proof line indicates that p is substituted for q in thesis 1, the second part indicates the substitution in thesis 2. Thus, applying modus ponens to the results of substitution we prove thesis 3.

Due to Łukasiewicz, it is well known that the implicational systems **K**, **S** and **K**, **B'**, **W** are equivalent [Hilbert und Bernays, 1968, ch. III.3].

It then follows that C_2 can be axiomatized with **K**, **B'**, **W**, and Ax.3. In the next chapter, we compare this axiomatization with the axiomatization of Łukasiewicz three-valued logic.

A logical calculus presented with the help of a set of axioms and inference rules is called a *Hilbert-style calculus*. If there is a proof of a formula A from a set of formulas Γ in a Hilbert-style calculus, we say that A *is a syntactic consequence of* Γ (symbolically, $\Gamma \vdash A$).

When constructing Hilbert-style proofs, it is very convenient to use an auxiliary rule called the rule of deduction; the validity of this rule is based on the following *deduction theorem*.

Deduction theorem. $\Gamma, A \vdash B$ *iff* $\Gamma \vdash A \supset B$. In particular, $A \vdash B$ *iff* $\vdash A \supset B$.

Starting off with a syntactic representation of a logic, we can identify it with the set of its theorems or its syntactic consequence relation. Thus, under a semantic approach, formulas are interpreted as functions over the set $\{0, 1\}$, while under a syntactic approach, formulas are nothing more than strings of symbols and we only distinguish theorems from not-theorems. That difference notwithstanding, both approaches to presenting the classical propositional logic are essentially equivalent or, what comes to the same thing, they are *adequate* with respect to each other. This means that the concepts of semantic and syntactic consequence extensionally coincide.

Adequacy theorem. $\vdash A$ *iff* $\vDash A$.

(a) If $\vdash A$ then $\vDash A$. This part of the adequacy theorem is called the *soundness theorem*. Soundness is a minimal condition we demand for a logical calculus. To prove it, all we have to do is check that all the axioms are tautologies and that the rules of inference preserve the property of being a tautology. It is easy to see, in this way, that C_2 is sound. Moreover, it is then easy to check that C_2 is consistent – for no formula A, both A and $\neg A$ can be theorems.

(b) If $\vDash A$ then $\vdash A$. This part of the adequacy theorem is called the (deductive, as opposed to functional) *completeness theorem*. This means that our axioms and rules of inference are sufficient for obtaining all the tautologies of C_2. The first published proof of this result belongs to Post ([Post, 1921]).

II.   Łukasiewicz Three-Valued Logic

II.1.   Logical fatalism

In the introduction to Łukasiewicz's *Selected Works* J. Słupecki stressed that "... the problem in which Łukasiewicz was most interested almost all his life and which he strove to solve with extraordinary effort and passion was the problem of determinism. It inspired him with the most brilliant idea, that of many-valued logics" [Łukasiewicz, 1970, p. vii].

The roots of Łukasiewicz many-valued logics can be traced back to Aristotle (4th century BC), who in the well known Chapter IX of his treatise *De Inrerpretatione* presents and refutes a fatalistic argument, which despite its intended simplicity, may appear baffling. The argument is as follows. An adherent of fatalism makes a move, in some way, from the *truth* of a proposition to the *necessity* of the event described by the proposition (*the principle of necessity*), and – in a similar vein – from the *falsity* of a proposition to the *impossibility* of the event described by it. So, if it is true that there will be a sea battle tomorrow, then it is necessary that there will be one, and if it is false that there will be a sea-battle tomorrow, then it is impossible that there will be one. More generally, since every proposition is either true or false (*the principle of bivalence*), then everything happens by necessity and there are neither contingencies nor, for that matter, any scope for free will.

Aristotle's argument can be formally stated as follows (compare with [McCall, 1968]):

1. $Tp \rightarrow Np$ premise (I)

2. $Fp \rightarrow N{\sim}p$ by analogy from (1)

3. $Tp \vee Fp$ premise (II)

4. $Np \vee N{\sim}p$ from (1), (2), and (3) by the rule of *complicated constructive dilemma*,

where 'T' stands for '*it is true that*'; 'F' stands for '*it is false that*'; 'N' – for '*it is necessary that*'; '$\rightarrow$' – for '*if ... then ...*'; '$\sim$' – for '*it's not the case that*', and '$\vee$' for '*or*'.

Note that we can obtain (2) by, first, using the usual definition of falsity $Fp \leftrightarrow T\sim p$, and then, applying *transitivity* to that formula and $T\sim p \rightarrow N\sim p$.[3]

Thus, we hit upon a link between the logical principle of bivalence and the predetermination of the future. The doctrine claiming that such a link really exists is called *the doctrine of logical fatalism* (*determinism*). Then, from the logical point of view, Aristotle's problem of future contingency boils down to solving the problem of the truth status of propositions about future contingent events. The discussion of Aristotle's problem of future contingency was already well under way in antiquity. Then, it resulted in a heated argument about free will and fatalism. In the Middle Ages that discussion was linked to the problem of God's omniscience. The problem is still being discussed, especially in connection with developments in modern logic (for more details, see [Jordan, 1963], [Karpenko, 1990]).

We will follow the so-called 'traditional' or 'standard' interpretation of Aristotle's solution of his problem; this interpretation was the one adhered to by Łukasiewicz. By way of refuting the fatalistic argument he presented, Aristotle emphasizes that propositions about *future contingent events* are neither actually true nor actually false. Hence the fatalistic argument fails since premise (II) is rejected. It is interesting that the premise (I) was accepted in the most Hellenistic philosophical schools (see [White, 1983]).

In 1920 Jan Łukasiewicz constructed a three-valued logic based on the metaphysics of 'indeterministic philosophy' (see [Łukasiewicz, 1970b]). The first mention of the three-valued logic was, however, made even earlier, in Łukasiewicz's lecture delivered in 1918. There, Łukasiewicz says: "that new logic ... destroys the former concept of science"; moreover, Łukasiewicz makes a connection between the "new logic" and the "struggle for the liberation of the human spirit" (see [Łukasiewicz, 1970a]). Philosophical ideas underlying the third truth-value are discussed in Łukasiewicz's seminal paper "On determinism" [Łukasiewicz, 1970c].[4] There, Łukasiewicz states that Aristotle's solution of the problem of future contingency destroys one of the main principles of our logic, namely that

---

[3] The modern version of Aristotle's fatalistic argument has been proffered by R. Taylor in [Taylor, 1962], which provoked an intense discussion well reviewed in [Bennett, 1974].

[4] This paper is a revised version of the address that Łukasiewicz delivered as a Rector at the start of the academic year 1922-23 at Warsaw University. Later on Łukasiewicz revised the address, giving it the form of a paper without changing the main claims and arguments. It was published for the first time in Polish in 1961. An English version of the paper was published in *Polish Logic 1920-1939* (S. Mccall ed.), Oxford, 1967, pp. 19-39, and also in [Łukasiewicz, 1970, pp. 110-128]. An up-to-date analysis of this paper and especially of Łukasiewicz's arguments against determinism is in [Becchi, 2002].

every proposition is either true or false. Łukasiewicz calls this principle *the principle of bivalence.*[5] According to him, it is an underlying principle of logic that can not be proved – one can only believe in it. Łukasiewicz claims that the principle of bivalence does not seem self-evident to him. Therefore, he claims to have the right not to accept it and to stipulate that, along with truth and falsity, there should be at least one more truth-value, which Łukasiewicz considers to be intermediate between the other two. Łukasiewicz concludes that "If this third value is introduced into logic we change its very foundations. A trivalent system of logic ... differs from ordinary bivalent logic, that only one known so far, as much as non-Euclidean systems of geometry differ from Euclidean geometry" [Łukasiewicz, 1970c, p. 126]. Similar passages occur in other papers of Łukasiewicz. Those claims foreshadowed a radical revision of the classical logic.

II.2.   Truth-tables. Axiomatization

The meaning of the third truth-value can be clarified by the following passage from Łukasiewicz: "I can assume without contradiction that my presence in Warsaw at a certain moment of next year, e.g., at noon on 21 December, is at the present time determined neither positively nor negatively. Hence it is *possible*, but not *necessary*, that I shall be present in Warsaw at the given time. On this assumption the proposition 'I shall be in Warsaw at noon on 21 December of next year', can at the present time be neither true nor false. For if it were true now, my future presence in Warsaw would have to be necessary, which is contradictory to the assumption. If it were false now, on the other hand, my future presence in Warsaw would have to be impossible, which is also contradictory to the assumption. Therefore, the proposition considered is at the moment *neither true nor false* and must posses a third value, different from '0' or falsity and '1' or truth. This value we can designate by '$1/2$'. It represents 'the possible', and joins 'the true' and 'the false' as a third value" [Łukasiewicz (1930), 1970e, pp. 165-166].

Adhering to the classical way of defining implication $p \rightarrow q$ and negation ~p wherever their arguments are the classical truth-values 0 and 1, Łukasiewicz defines the meaning of those connectives for the cases featuring his new truth-value in the following way:

---

[5] See also Appendix "On the history of the law of bivalence" in [Łukasiewicz (1930), 1970e].

$$(1 \to {}^1/_2) = ({}^1/_2 \to 0) = {}^1/_2,$$
$$(0 \to {}^1/_2) = ({}^1/_2 \to {}^1/_2) = ({}^1/_2 \to 1) = 1,$$
$$\sim {}^1/_2 = {}^1/_2.$$

The other propositional connectives are defined by means of the primary connectives:

$$p \lor q = (p \to q) \to q \qquad \text{(disjunction)}$$
$$p \land q = \sim(\sim p \lor \sim q) \qquad \text{(conjunction)}$$
$$p \leftrightarrow q = (p \to q) \land (q \to p) \qquad \text{(equivalence)}.$$

Thus, the truth-tables for the logical connectives look as follows:

p	$\sim p$
1	0
${}^1/_2$	${}^1/_2$
0	1

$\to$	1	${}^1/_2$	0
1	1	${}^1/_2$	0
${}^1/_2$	1	1	${}^1/_2$
0	1	1	1

$\lor$	1	${}^1/_2$	0
1	1	1	1
${}^1/_2$	1	${}^1/_2$	${}^1/_2$
0	1	${}^1/_2$	0

$\land$	1	${}^1/_2$	0
1	1	${}^1/_2$	0
${}^1/_2$	${}^1/_2$	${}^1/_2$	0
0	0	0	0

$\leftrightarrow$	1	${}^1/_2$	0
1	1	${}^1/_2$	0
${}^1/_2$	${}^1/_2$	1	${}^1/_2$
0	0	${}^1/_2$	1

A *valuation* of the set *For* is a function v: *For* $\to \{0, {}^1/_2, 1\}$, 'compatible' with the above truth-tables. A *tautology* is a formula which under any valuation v takes on the *designated value* 1. The set of thus defined tautologies is *Łukasiewicz three-valued logic* $Ł_3$.

In 1931, M. Wajsberg showed that the three-valued Łukasiewicz logic can be axiomatized in the following way ([Wajsberg, 1977a]):

1. $(p \to q) \to ((q \to r) \to (p \to r))$

2. $p \to (q \to p)$

3. $(\sim p \to \sim q) \to (q \to p)$

4. $((p \to \sim p) \to p) \to p$.

The rules of inference are as in the classical propositional logic C_2.

R1. *Modus ponens*: If A and $A \to B$, then B.

R2. *Substitution*: If $\vdash A(p)$, then $\vdash A(B)$, where B is a well-formed formula that is substituted uniformly for p.

Thus, the axiomatization of $Ł_3$ is obtained from the axiomatization of C_2 with axioms **B'**, **K**, **W**, and inverse contraposition (see section I.4) by replacing **W** with Wajsberg's axiom (4). It is, however, easy check that a weakened form of **W**, formula $\mathbf{W_1}$:

$$(p \to (p \to (p \to q))) \to (p \to (p \to q))$$

is a tautology of $Ł_3$.

Wajsberg's axiomatization means that for $Ł_3$ the following theorem holds:

Adequacy theorem. $\vdash A$ *iff* $\vDash A$.[6]

Thus, $Ł_3$ is, like C_2, deductively complete and consistent.

II.3. Differences between $Ł_3$ and C_2

Its worth pointing out again that the behavior of the connectives of $Ł_3$ over the set $\{1, 0\}$ coincides with that of the connectives of C_2. Thus, logic $Ł_2$ (the two-valued version of $Ł_3$) is nothing else but the classical propositional logic C_2. It is evident that any tautology of $Ł_3$ is a tautology of C_2, but not *vice versa*.

We already know that the *contraction law*

$$(p \to (p \to q)) \to (p \to q)$$

is not a tautology of $Ł_3$ (assign $^1/_2$ to p and 0 to q, to get a counterexample).[7]

In fact, $Ł_3$ is radically different from C_2 – some important laws of the classical logic, such as

$$p \vee {\sim}p \qquad \text{(the law of the excluded middle)}$$

$$\sim(p \wedge {\sim}p) \qquad \text{(the law of non-contradiction)}$$

fail in $Ł_3$ (as well as in any $Ł_n$): these formulas take on the value $^1/_2$ when p is $^1/_2$.

It is worth mentioning that $Ł_3$ has been much criticized for failing the law of non-contradiction (for references, see [Karpenko, 1982]).

---

[6] In [Epstein, 1990] adequacy theorem for $Ł_3$ is proved in the form в виде $\Gamma \vdash A$ *iff* $\Gamma \vDash A$.

[7] So $Ł_3$ is a '*resource conscious*' logic; it is also historically the first *logic without contraction*. [Ono and Komori, 1985] in which Łukasiewicz's logics are discussed from such point of view.

If, however, we consider both 1 and $^1/_2$ as designated truth-values of $Ł_3$, then $p \vee \sim p$ and $\sim(p \wedge \sim p)$ become tautologies. It was widely believed (this belief is amply present in the literature) that no two-valued tautology can take on the value 0 in $Ł_3$ until R.A. Turquette found the formula

$$\sim(p \rightarrow \sim p) \vee \sim(\sim p \rightarrow p),$$

which is a tautology in C_2, but takes on the value 0 in $Ł_3$ when p is $^1/_2$. (Note that Turquette's formula is equivalent to $\sim(p \leftrightarrow \sim p)$.)

For us, the most important difference between $Ł_3$ and C_2 is the following. In 1936, J. Słupecki showed that $Ł_3$, by contrast with C_2, is not truth-functionally complete; that is, not every three-valued truth-function can be defined in $Ł_3$. To see this, take Słupecki's operator Tp:

p	Tp
1	$^1/_2$
$^1/_2$	$^1/_2$
0	$^1/_2$

Tp can not be defined in $Ł_3$; furthermore, by adding Tp to $Ł_3$, we get a truth-functionally complete system. Further on in the book (section IV.10), we will discuss the *functional properties* of $Ł_3$ in more detail. The functional properties of the generalizations of $Ł_3$ to an arbitrary number n ($n \geq 3$, $n \in N$) of truth-values will prove crucial for our investigation.

It is worth noting that, unlike the classical logic, $Ł_3$ is rich enough to provide the means for defining two *non-trivial* truth-functional modal operators ◊ and □. The following definition of possibility was suggested by A. Tarski in 1921 (see [Łukasiewicz, 1970d, p. 167]):

$$◊p = \sim p \rightarrow p.$$

Necessity is then defined as usual:

$$□p = \sim ◊ \sim p.$$

The truth-tables for these operators look then as follows:

p	◊p	□p
1	1	1
$^1/_2$	1	0
0	0	0

Then, $\nvDash (p \to \Box p)$ in $Ł_3$, but $\vDash (p \to (p \to \Box p))$.

It is worth noting that we can not define Łukasiewicz implication $\to$ through $\sim$, $\vee$, and $\wedge$ alone[8], but if we add Tarski's modal operators to those connectives, then the following definition due to J. Słupecki does the trick (it was suggested in 1964 in an attempt to give an intuitive interpretation of $Ł_3$ – see [Słupecki, Bryll and Prucnal, 1967, p. 51]):

$$p \to q = (\sim p \vee q) \vee \Diamond(\sim p \wedge q).$$

As remarked by Słupecki, the above interpretation of the implication complies with our intuition.

An axiomatization of $Ł_3$ with the help of the characteristic modal schemas of **S5** and a resultant considerable simplification of Wajsberg's ingenious completeness proof can be found in [Minari, 1991]. It is also worth taking a look at an embedding of $Ł_3$ into **S5** suggested in [Woodruff, 1974].

II.4. An embedding C_2 into $Ł_3$ and three-valued isomorphs of C_2.

An immediate consequence of the failure of the contraction law in $Ł_3$ is that the standard form of the *deduction theorem* does not hold for $Ł_3$ since

$$p \wedge \sim p \vDash q \text{ but } \nvDash (p \wedge \sim p) \to q.$$

The connective appropriate for some form of the deduction theorem can be the following:

$$p \to_1 q = p \to (p \to q).$$

Its truth-table is

$\to_1$	1	$^1/_2$	0
1	1	$^1/_2$	0
$^1/_2$	1	1	1
0	1	1	1

---

[8] The logic that has exactly these connectives as its initial connectives is *Kleene's three-valued logic* [Kleene, 1952, § 64].

The truth-table for $\rightarrow_1$ was independently suggested, on the one hand, by Słupecki, Bryll, and Prucnal in the already mentioned paper, and on the other, by A. Monteiro in [Monteiro, 1967]. In the former work, $p \rightarrow_1 q$ was defined as $\sim\Box p \vee q$ and an axiomatization of $\mathbf{Ł_3}$ was given with the help of the connectives $\sim$, $\Box$ and $\vee$. In the latter work, $p \rightarrow_1 q$ was defined as $\Diamond\sim p \vee q$ and it was noticed that we could take $\sim$, $\wedge$ and $\rightarrow_1$ as the primary connectives for $\mathbf{Ł_3}$, defining $p \rightarrow q$ as

$$(p \rightarrow_1 q) \wedge (\sim q \rightarrow_1 \sim p).$$

The advantage of the above-given definition of $p \rightarrow_1 q$ is, however, that it easily generalizes to Łukasiewicz n-valued logics $\mathbf{Ł_n}$ (see below) by iterating the '$p \rightarrow$' part of the definition (see [Wójcicki, 1988, p. 72]).

As has already been mentioned, $\mathbf{Ł_3} \subset \mathbf{C_2}$. But we can show that $\mathbf{C_2}$ is, in a sense, richer than $\mathbf{Ł_3}$. Moreover, $\mathbf{Ł_3}$ is, in a sense, an extension $\mathbf{C_2}$! First let's consider an *embedding* (translation) of $\mathbf{C_2}$ into $\mathbf{Ł_3}$ based on the ideas of M. Tokarz [Tokarz, 1971].

The embedding operation * (the map) is given by:

$$(p)^* = p,$$

$$(A \supset B)^* = A^* \rightarrow_1 B^*,$$

$$(\neg A)^* = A^* \rightarrow_1 \sim(A^* \rightarrow A^*).$$

Theorem 1. $\vDash A$ *in* $\mathbf{C_2}$ *iff* $\vDash A^*$ *in* $\mathbf{Ł_3}$ (see [Epstein, 1990, p. 238]).

We can suggest another embedding operation φ:

$$\varphi(p) = \Box p,$$

$$\varphi(A \supset B) = \varphi(A) \rightarrow \varphi(B),$$

$$\varphi(\neg A) = \sim\varphi(A),$$

where $\Box$ is the above defined Tarski's operator of necessity.

Theorem 2. $\vDash A$ *in* $\mathbf{C_2}$ *iff* $\vDash \varphi(A)$ *in* $\mathbf{Ł_3}$.

This embedding suggests the following definitions of two new connectives:

$$p \rightarrow^\Box q = \Box p \rightarrow \Box q,$$

$$\sim^\Box p = \sim\Box p.$$

The truth-tables for these connectives are

p	$\sim^{\square}p$
1	0
$1/2$	1
0	1

$\rightarrow^{\square}$	1	$1/2$	0
1	1	0	0
$1/2$	1	1	1
0	1	1	1

Note that we could take $\sim$, $\wedge$, $\vee$ and $\rightarrow^{\square}$ as the initial connectives for $\textbf{Ł}_3$, since

$$p \rightarrow_1 q = (p \rightarrow^{\square} q) \vee q, \text{ and } p \rightarrow q = (p \rightarrow_1 q) \wedge (\sim q \rightarrow_1 \sim p).$$

The axiomatization of $\textbf{Ł}_3$ with $\sim$, $\wedge$, $\vee$ and $\rightarrow^{\square}$ as initial connectives was set out by V.K. Finn in [Finn, 1974, p. 426-427].

Let us denote the logic with the connectives $\sim^{\square}$ and $\rightarrow^{\square}$ as $\textbf{Ł}_3^{\square}$. This logic was first constructed by the Russian logician D.A. Boczvar in 1938 (see [Boczvar, 1981]), who gave it the name of a '*three-valued isomorph of classical propositional logic*'. By such a name, he meant to convey that that the rule of modus ponens is valid in $\textbf{Ł}_3^{\square}$ and that $\vDash A$ *in* $\textbf{C}_2$ iff for every $\textbf{Ł}_3^{\square}$-evaluation e, $e(A) = 1$ (see also [Resher, 1969, p. 32]).

In 1999, a student of the author's, E. Komendantsky wrote a computer program that calculated that Łukasiewicz three-valued logic $\textbf{Ł}_3$ contains 18 implication-negation fragments isomorphic to $\textbf{C}_2$. Among them only 2 isomorphs, namely $\{\sim^{\square}, \rightarrow^{\square}\}$ and $\{\sim^{\square}, \rightarrow_1\}$, have one designated value 1. For most of the isomorphs, a second designated value is needed to validate modus ponens. At the end of the day, however, we have 18 embedding operations embedding $\textbf{C}_2$ into $\textbf{Ł}_3$. For example, the embedding operation ψ which gave an isomorph with two designated values is the following:

$$\psi(p) = \Diamond p,$$
$$\psi(A \supset B) = \psi(A) \rightarrow \psi(B),$$
$$\psi(\neg A) = \sim \psi(A),$$

where $\Diamond$ is the above defined Tarski's operator of possibility.

Theorem 3. $\vDash A$ *in* $\mathbf{C_2}$ *iff* $\vDash \psi(A)$ *in* $\mathbf{Ł_3}$.

This embedding suggests the following definition of two new connectives:

$$p \to^\Diamond q = \Diamond p \to \Diamond q,$$

$$\sim^\Diamond p = \sim\Diamond p.$$

The truth-tables for these connectives are:

p	$\sim^\Diamond p$
1	0
$^1/_2$	0
0	1

$\to^\Diamond$	1	$^1/_2$	0
1	1	1	0
$^1/_2$	1	1	0
0	1	1	1

The existence within $\mathbf{Ł_3}$ of isomorphs of $\mathbf{C_2}$ hints at the possibility to axiomatize $\mathbf{Ł_3}$ as an extension of $\mathbf{C_2}$. For $\mathbf{Ł_3}$ (with two designated values) this was done by I.M.L. D'Ottaviano and R. Epstein in [Epstein, 1990, p. 279]. For a very large class of n-valued logic, including all finite-valued Łukasiewicz logics $\mathbf{Ł_n}$, such an axiomatization for logics with an arbitrary number of designated values was suggested by O. Anshakov and S. Rychkov in 1982 (see [Anshakov and Rychkov, 1994]).

Thus, $\mathbf{Ł_3}$ as well as all $\mathbf{Ł_n}$'s, can be axiomatized as an extension of the classical propositional calculus.

III. Łukasiewicz finite-valued logics

III.1. Łukasiewicz n-valued matrix

The generalization of his three-valued logic led Łukasiewicz, in 1922, to consider the n-valued logic ([Łukasiewicz, 1970d]). The technical results concerning this logic appeared in the famous compendium of 1930 (see [Łukasiewicz and Tarski, 1970]).

A logical matrix of the form

$$\mathfrak{M}_n^L = <V_n, \sim, \rightarrow, \{1\}>$$

is called a *Łukasiewicz n-valued matrix* $(n \in N, n \geq 2)$ provided that

$$V_n = \{0, 1/n\text{-}1, ..., n\text{-}2/n\text{-}1, 1\};$$

$\sim$ (negation) is a unary function, and $\rightarrow$ (implication) is a binary function; these are defined on V_n as follows:

$$\sim x = 1 - x,$$
$$x \rightarrow y = min(1, 1 - x + y);$$

$\{1\}$ is the set of the designated elements of $\mathfrak{M}_n^L$.

Functions $\vee$ (disjunction), $\wedge$ (conjunction), and $\equiv$ (equivalence) are defined through the above-mentioned operations as follows:

$$x \vee y = (x \rightarrow y) \rightarrow y = max(x, y)$$
$$x \wedge y = \sim(\sim x \vee \sim y) = min(x, y)$$
$$x \equiv y = (x \rightarrow y) \wedge (y \rightarrow x).$$

III.2. Matrix logic Ł_n

To the algebras $<V_n, \sim, \rightarrow>$ $(n \geq 2)$ of the matrix $\mathfrak{M}_n^L$ in the usual way there corresponds propositional language *SL* containing an infinite stock of propositional variables $p, q, r, ..., p_1, q_1, r_1, ...$ and two connectives: $\sim$ (negation) and $\rightarrow$ (implication). From an algebraic point of view, language *SL* has an algebraic structure which is the structure of an absolutely free algebra of type $<1, 2>$. This enables us to define *valuations* of *SL* into $\mathfrak{M}_n^L$ as homomorphisms from *SL* into $\mathfrak{M}_n^L$ (strictly speaking, into $<M_n, \sim, \rightarrow>$).

Lastly, we define Łukasiewicz n-valued matrix logic $Ł_n$ to be the set of all tautologies of the matrix $\mathfrak{M}_n^L$, i.e. the set of all such formulas A that $v(A) = 1$ for each valuation v of SL into $\mathfrak{M}_n^L$.

The problem of interrelations among the finite-valued systems $Ł_n$ was settled by A. Lindenbaum (see [Łukasiewicz and Tarski, 1970, p. 142]) in the following way:

$$Ł_m \subseteq Ł_n \text{ if and only if } n\text{-}1 \text{ divides } m\text{-}1.^9$$

III.3.   Axiomatization of $Ł_n$

It is not straightforward to find a finite axiomatization of the set of tautologies of $Ł_n$ for arbitrary n. Although the axiom system given by Wajsberg (see section I.4) was simple, but there was no hint of how to extend this method to the other systems. It is claimed that A. Lindenbaum proved finite axiomatizability of all n-valued logics, where n-1 is a prime number (!), however this result has never been published (see [Tuziak, 1988, p. 49]). Later, a general theorem on finite axiomatizability of a large class of finite logics, including all $Ł_n$, was given by Wajsberg in 1935 (see [Wajsberg, 1977]).

The problem of axiomatizing namely $Ł_n$ for arbitrary n was solved only in 1952 by Rosser and Turquette (see [Rosser and Turquette 1952]), but their method, say nothing of Wajsberg's method, has no practical value. Rather simple axiomatization of $Ł_n$ was suggested in 1973 by R. Grigolia, using algebraic methods (see [Grigolia, 1977]). A more sophisticated axiomatization of $Ł_n$ was suggested Tokarz in [Tokarz, 1974a]. (In both cases, completeness for finite-valued logics is derived from the completeness of the infinite-valued Łukasiewicz logic.) An axiomatization presenting $Ł_n$ as an extension of the intuitionistic propositional calculus was given by Cignoli in [Cignoli, 1982]. Lastly, a very simple axiomatization was suggested by Tuziak in [Tuziak, 1988]; this is the axiomatization we consider in what follows.

We use the following abbreviations: $p \to^0 q = q$, $p \to^{k+1} q = p \to (p \to^k q)$ and $p \equiv q = (p \to q) \wedge (q \to p)$. The axioms of $Ł_n$ are

---

[9] See the proof in [Uquhart, 1986, pp. 83-84].

1. $(p \to q) \to ((q \to r) \to (p \to r))$.
2. $p \to (q \to p)$.
3. $((p \to q) \to q) \to ((q \to p) \to p)$.
4. $(p \to^n q) \to (p \to^{n-1} q)$.
5. $p \wedge q \to p$.
6. $p \wedge q \to q$.
7. $(p \to q) \to ((p \to r) \to (p \to q \wedge r))$.
8. $p \to p \vee q$.
9. $q \to p \vee q$.
10. $(p \to r) \to ((q \to r) \to (p \vee q \to r))$.
11. $(\sim p \to \sim q) \to (q \to p)$.
12. $(p \equiv (p \to^{s-2} \sim p)) \to^{n-1} p$ for any $2 \leq s \leq n-1$ such that s is not a divisor of $n-1$.

The rules of inference are *modus ponens* and *substitution*.

Let's consider some examples. When $n = 2$ or $n = 3$, formulas (1) – (11) are the only axioms; there are no axioms of the form (12). It is worth noting that, in the case of $n = 2$, we get the set of axioms for the classical propositional logic C_2, axiom (4) being "the law of contraction":

$$(p \to (p \to q)) \to (p \to q).$$

When $n = 3$, (4) becomes

$$(p \to (p \to (p \to q))) \to (p \to (p \to q)).$$

The last formula is the only "new"axiom.

If $n = 4$, then we have only one axiom of the form (12):

$$(p \equiv \sim p) \to ((p \equiv \sim p) \to ((p \equiv \sim p) \to p)).$$

Note that it suffices to consider only primes s in (12).

As mentioned above, Łukasiewicz logic $Ł_n$ is, historically, the first logic without the law of contraction; furthermore, adding this formula to $Ł_n$ ($n \geq 3$) immediately gives us C_2. It is also worth mentioning the paper [Prijatelj, 1996], where Gentzen-style calculi for $Ł_n$, based on the restriction of the structural rule of contraction, are presented.

III.4. Cardinal degree of completeness of $Ł_n$

A. Tarski in [Tarski, 1930b] introduced the notion of the cardinal degree of completeness of a logic. This notion may be defined in the following way.

Let L be a logic. The cardinal degree of completeness of L, in symbols $\gamma(L)$, is the number of logics containing the theorems of L.

Tarski proved (see [Łukasiewicz and Taski, 1970, p. 142]) that

$\gamma(Ł_n) = 3$ *for $n-1$ being a prime number.*

Let $C = \langle a_1, \ldots, a_n\rangle$ be an arbitrary sequence of natural numbers. We denote by $N_c(a_i)$ ($1 \le i \le n$) the number of subsequences D of C which satisfy the following condition:

$a_i \in D$ and for every $b \in D$, $a_i \ge b$,
if $j \ne k$ and $a_j, a_k \in D$, then a_j-1 is not a divisor of a_k-1.

Let $c(n) = \langle a_1, \ldots, a_k\rangle$ be the sequence that has the following properties:
  (i) $a_1 = n$,
  (ii) $a_1 > \cdots > a_k > 1$,
  (iii)   for every i, $1 \le i \le k$, a_i-1 is a divisor of $n-1$.

Theorem 1. For finite n,

$$\gamma(Ł_n) = \left(\sum_{a_i \in c(n)} N_{c(n)}(a_i)\right) + 1.$$

The proof of Theorem 1 was first published by M. Tokarz in [Tokarz, 1974b]; a shorter proof can be found in [Tokarz, 1977].

In 2000, M.N. Rybakov (Tver State University, Russia) wrote a computer program for calculating $\gamma(Ł_n)$. Table 1, which appears to be of great interest, contains the values of $\gamma(Ł_n)$ for $n \le 1000$. Apparently, some natural numbers are not values of $\gamma(Ł_n)$ for any n. So, for the first ten thousand n, the values of $\gamma(Ł_n)$ contain the following natural numbers from the first one hundred:

2, 3, 4, 5, 6, 7, 8, 9, 10, 11, 12, 14, 15, 20, 21,
28, 35, 36, 45, 50, 55, 56, 66, 70, 78, 84, 91.

III.5. Interpretations of $Ł_n$

Algebraic semantics for $Ł_n$ may be found in [Grigolia, 1977] and [Cignoli, 1982]. R. Grigolia developed some ideas of C.C. Chang (see [Chang, 1958, 1959]), who in his algebraic completeness proof of the infinite-valued Łukasiewicz logic $Ł_\infty$ (where the set of truth-values is the interval $[0,1]$), introduced the notion of MV-algebra and proved the representation theorem for such algebras.[10] MV_n-algebras are not based on lattices; thus, their comparison with algebraic structures corresponding to other logical calculi is not straightforward. In the semantics given by Cignoli, the n-valued Łukasiewicz algebras MV_n are based on algebras introduced in 1940 by G. Moisil; these are bound distributive lattices on which a de Morgan negation and $n-1$ modal operators (more specifically, these modal operators are Boolean-valued endomorphisms) are defined. Cignoli showed that adequate algebraic counterparts for $Ł_n$ can be obtained by adding to the Moisil algebras a set of $(n(n-5)+2)/2$ $(n \geq 5)$ binary operators satisfying simple equations.

The main problem, however, with the interpretation of $Ł_n$ is that of the interpretation of it truth-values. Despite much progress made in the study of Łukasiewicz n-valued logics (see [Wójcicki and Malinowski (eds.), 1977], and the bibliography on the main applications of $Ł_n$ compiled by Malinowski in this book), the interpretation of their truth-values is it still a cause of much controversy. Thus, Dana Scott has once said: "Before you accept many-valued logic as a long-lost brother, try to think what these fractional truth values could possibly mean. And do they have any use? What is the conceptual justification of 'intermediate' values?" ([Scott, 1976]). The substantiation of logical operations of $Ł_n$ (especially, implication) is not clear, either.

G. Malinowski in his monograph "Many-valued logics" gave three interpretations of $Ł_n$ in terms zero-one (falsity-truth) valuations [Malinowski, 1993, ch. 10]: Suszko's thesis, Scott's method, and Urquhart's interpretation.

The first of these provides a *valuation semantics*. The bivaluations, considered as characteristic functions of sets of formulas, were introduced independently by N.C.A. da Costa (for paraconsistent logics), D. Scott, and R. Suszko in the early 70s. In their works, these authors clearly distinguished bivaluations from the more restricted class of algebraic valuations, which are homomorphisms between abstract algebras (see section 1.2). So, a two-valued *logical valuation* is simply a function which associates one value to each formula. A semantics based on a logical valuation is called a *valuation semantics*, in contrast to semantics based on algebraic valuations. Suszko

---

[10] In details see [Cignoli, D'Ottaviano and Mundici, 2000].

([Suszko, 1975]) gave the valuation semantics for Łukasiewicz three-valued logic $Ł_3$[11] and claimed that any many-valued logic is a two-valued logic in disguise! Of course, it is not quite true: the use of two truth-values does not automatically make a logic two-valued: in addition, valuations should be homomorphisms. For detailed discussion, see [da Costa, Béziau and Bueno, 1996], [Tsuji, 1998], and especially [Caleiro, Carnielli, Coniglio and Marcos, 2005].

We will see some of the fundamental differences between the classical two-valued logic and three-valued logics in the next chapter. At this point, it is worth noticing that every propositional logic has a valuation semantics. There are several ways to show that any logic may be given semantics with only two truth-values (see, for example, [Routley and Meyer, 1976]). (We are here interested, however, in distinguishing features of $Ł_n$. rather than in its similarities with other logics.) Now, we describe briefly the other two above-mentioned approaches to interpreting the truth-values of $Ł_n$.

Scott [Scott, 1974] interprets the elements of a model for $Ł_n$ as representing *degrees of error*. Scott suggests that '$v_i(A) = t$' for $i \in \{0, ..., n\}$, should be read as '*the statement A is true to within degree of error i*'. Scott assumes that numbers in the range $0 \leq i \leq n$ stand for *degrees of error in derivation from the truth*: degree 0 represents no error at all (the truth), while the higher elements of a model represent greater degrees of error. Under this interpretation, all the tautologies of Łukasiewicz logics are schemas of the statements with error degree 0.

The negation and implication are then characterized in the following way for any $i, j, k \in (0, ..., n\}$:

$(\sim)$ $v_k(\sim A) = t$ if and only if $v_{n-k}(A) = f$

$(\rightarrow)$ $v_k(A \rightarrow B) = t$ if and only if whenever $i+k \leq j$ and $v_i(A) = t$, $v_j(B) = t$.

Smiley [Smiley, 1976], however, in a comment on another paper of Scott's [Scott, 1976], points out some difficulties in this interpretation. At the beginning of [Scott, 1976], Scott remarks that the probability of $(p \rightarrow q)$ cannot be a function of the probabilities of p and q; the same must be true for the logic of error. Thus, A. Urquhart asserts that "The logic of uncertainty, the logic of probability and the logic of error are all non-truth-functional" [Urquhart, 1986, p. 106], whereas Łukasiewicz logics, as well as the classical two-valued logic, are truth-functional, i.e. the values of complex propositions are determined by the values of their components.

---

[11] G. Malinowski [Malinowski, 1977] presented the *valuation semantics* for $Ł_n$.

Similar to Scott's is Urquhart's interpretation of $\mathbf{L_n}$ (this similarity was noted by Scott himself). Urquhart [Urquhart, 1973] suggested Kripke-style semantics for $\mathbf{L_n}$, with elements $\{0, ..., n\}$ of a model being time instants rather than degrees of error. Of course, at every time instant a proposition is either true or false. However this time interpretation is criticized in [Rine, 1974].

In the late 70s, two other interpretations of $\mathbf{L_n}$ were suggested. M. Byrd presented an interpretation of truth-values of $\mathbf{L_n}$ in terms of *T-F-sequences*, where T is the truth and F is falsity [Byrd, 1979]. At around the same time, and independently from Byrd, A.S. Karpenko suggested an interpretation of truth-values of $\mathbf{L_n}$ in terms of *sets* of T-F-sequences (see [Karpenko, 1983]).

III.5.1. T-F-sequences as truth-values

First, let's introduce some notions. Let $B = \{T, F\}$ be the set of classical truth-values. Then, for any natural number $s \geq 2$

$$B^S = \{<a_1, ..., a_s> \mid a_i \in B\}, 1 \leq i \leq s.$$

The elements of B^S, i.e. *T-F-sequences*, are designated as α, β, γ with or without indices.

$$\text{Algebra } \mathcal{A}_s^B = < B^S, \neg^+, \supset^+, \vee^+, \wedge^+ >$$

is a Boolean algebra with 2^S elements, where the operations $\neg^+, \supset^+, \vee^+, \wedge^+$ are defined component-wise through the Boolean operations $\neg, \supset, \vee, \wedge,$ in the following way: for any *T-F-sequences* $\alpha = <a_1, ..., a_s>$ и $\beta = <b_1, ..., b_s>$

$$\neg^+\alpha = <\neg a_1, ... , \neg a_s>,$$

$$\alpha \supset^+ \beta = <a_1 \supset b_1>, ... , <a_s \supset b_s>,$$

$$\alpha \vee^+ \beta = <a_1 \vee b_1>, ... , <a_s \vee b_s>,$$

$$\alpha \wedge^+ \beta = <a_1 \wedge b_1>, ... , <a_s \wedge b_s>.$$

Now, let's consider Byrd's interpretation of $\mathbf{L_n}$, which is of particular interest here since Byrd gives component-wise interpretations of operations of Łukasiewicz matrix $\mathfrak{M}_n^L$, unlike the interpretation of negation in Post's n-valued logic (see section II.5). To that end, however, Byrd introduces the one-place operation $d(\alpha)$, which transforms *T-F-sequences* so that all occurrences of T shift to the start of the sequence:

$$d(\alpha) = <T,T, ..., T,F,F, ..., F>.$$

Under our approach, this looks as follows. Let's consider the logical matrix

$$\mathfrak{M}^L_{s+1} = <B^S_T, d, \neg^d, \rightarrow^d, \{T^S\}>,$$

where B^S_T is the set of truth-values comprising only those T-F-sequences in which all occurrences of T are at the start of the sequence. It is easy to see that the number of such T-F-sequences is $s + 1$ and is equal to the number of truth-values of $\mathbf{L_n}$, i.e. $n = s + 1$. The elements of B^S_T will be designated as $\alpha^T, \beta^T, \gamma^T ...$. $\{T^S\}$ is a one-element set of designated elements where T^S is $< T,T, ..., T >$ and the operations are defined as follows:

1. $d(\alpha) = \alpha^T$.

2. $\neg^d(\alpha^T) = d(\neg^+(\alpha^T))$.

3. $\alpha^T \rightarrow^d \beta^T = d(\alpha^T \supset^+ \beta^T)$.

Theorem 2. *Matrices* $\mathfrak{M}^L_n = <M_n, \sim, \rightarrow, \{1\}>$ *and* $M^L_{s+1} = <B^S_T, d, \neg^d, \rightarrow^d, \{T^S\}>$ *are isomorphic.*

Hence $\mathfrak{M}^L_{s+1}$ is a characteristic matrix for logic $\mathbf{L_n}$, i.e. a formula A is a theorem of $\mathbf{L_n}$ if and only if A is a tautology of $\mathfrak{M}^L_{s+1}$.

Note that if the operation $d(\alpha)$ transforms T-F-sequences so that all occurrences of F are at the beginning:

$$d(\alpha) = < F,F, ..., F,T,T, ..., T>,$$

then we again obtain a matrix isomorphic to $\mathfrak{M}^L_{s+1}$. Thus, interpretation of $\mathbf{L_n}$ does not depend on whether the occurrences of F stand at the beginning or at the end of T-F-sequences.

Thus, we can see that $\mathbf{L_n}$ can be interpreted using different types of T-F-sequences (for example, 1/3 might be interpreted as $<T,F,F>$ or as $<F,F,T>$). It is, therefore, worth considering a semantics that does not restrict the choice of T-F-sequences as truth-values and where initial operations are interpreted as well-known Boolean component-wise operations, without the additional operator $d(\alpha)$. Such a semantics can be obtained through the structuralization of truth-values themselves: instead of T-F-sequences of a definite type, certain sets of T-F-sequences are taken as truth-values. This semantics is called 'factor-semantics' (see [Karpenko, 1983]; the factor-semantics with infinite T-F-sequences is presented in [Karpenko, 1988]). As was noted by the author ([Karpenko, 2000, p. 272]), such semantics is adequate for an extension of $\mathbf{L}_\infty$ that is the only pre-tabular extension of $\mathbf{L}_\infty$ (see [Beavers, 1993]).

III.5.2. Factor-semantics for $\mathbf{L}_n$

Let s be a natural number such that $s \geq 2$. Let's consider the algebra

$$\mathcal{A}_s = < B^s, \cong, R, \neg^+, \supset^+ >,$$

where B^s, $\neg^+$, and $\supset^+$ are defined as above. So, the Boolean algebra $< B^s, \neg^+, \supset^+ >$ is a semantic foundation for $\mathbf{L}_n$.

Let $\eta_T(\alpha)$ be the number of occurrences of T in α. Then $\alpha \cong \beta$ iff $\eta_T(\alpha) = \eta_T(\beta)$, and R is defined by the following way:

$$< a_1, ..., a_s > R < b_1, ..., b_s > \textit{ iff}$$

$$\begin{cases} \eta_T(\alpha) \leq \eta_T(\beta) \ \& \ \forall i \leq s \ (a_i = T \Rightarrow b_i = T) \textit{ or} \\ \eta_T(\alpha) > \eta_T(\beta) \ \& \ \forall i \leq s \ (b_i = T \Rightarrow a_i = T). \end{cases}$$

The relation R is reflexive and symmetrical, but generally, not transitive.

Definition 1. We say that a matrix $\mathfrak{N}^L_{s+1}$ is associated with the algebraic system $\mathcal{A}_s$ iff

$$\mathfrak{N}^L_{s+1} = < B^s/\cong, \neg^*, \rightarrow^*, \{|T^s|\} >, \text{ where}$$

1. $B^s/\cong$ is the factor set of B^s with respect to $/\cong$. Of course, $B^s/\cong$ has $s+1$ elements. If $\alpha \in B^s$, then $|\alpha|$ will designate the equivalence class of α.

2. The set of designated elements of $\mathfrak{N}^L_{s+1}$ is the one-element set containing only $|T^s|$.

3. For $|\alpha|, |\beta| \in B^s/\cong$ we put $\neg^*|\alpha| = |\neg^+\alpha|$ и $|\alpha| \rightarrow^* |\beta| = |\alpha' \supset^+ \beta|$, where $\alpha' \in |\alpha|$, $\beta \in |\beta|$ and $\alpha' R \beta$.

The above definition of $\mathfrak{N}^L_{s+1}$ is consistent since we have

Lemma 1. *The above-mentioned definitions of $\neg^*$ and $\rightarrow^*$ are sound.*

The proof of Lemma 1 is straightforward, so we leave it out.

Theorem 3. *Matrix $\mathfrak{N}^L_{s+1} = < B^s/\cong, \neg^*, \rightarrow^*, \{|T^s|\} >$ is adequate for Łukasiewicz n-valued logic $\mathbf{L}_n$, where $n = s + 1$.*

Proof. It suffices to show that the matrix $\mathfrak{N}^{L}_{s+1}$ and the Łukasiewicz matrix $\mathfrak{M}^{L}_{n}$ are isomorphic. The required isomorphism may be obtained by the mapping φ such that, for any $|\alpha| \in B^{S}/\cong$, $\varphi(|\alpha|) = \frac{\eta(\alpha)}{s}$.

It is obvious that φ is one-to-one and onto. We, thus, only have show that

$$(*) \quad \varphi(\neg^{*}|\alpha|) = \sim\varphi(|\alpha|),$$

$$(**) \quad \varphi(|\alpha| \to^{*} |\beta|) = \varphi(|\alpha|) \to \varphi(|\beta|).$$

The following proves (*):

$$\varphi(\neg^{*}|\alpha|) = \varphi(|\neg^{+}\alpha|) = \frac{s - \eta_{T}(\alpha)}{s} = 1 - \frac{\eta_{T}(\alpha)}{s} = 1 - \varphi(|\alpha|) = \sim \varphi(|\alpha|).$$

To prove (**), let's take $\alpha' \in |\alpha|$ and $\beta' \in |\beta|$, where $\alpha' R \beta'$. There are two cases to consider:

1. $\eta_{T}(\alpha) \leq \eta_{T}(\beta)$. Then, obviously, the right-hand side of (**) is 1. Furthermore, $|\alpha| \to^{*} |\beta| = |\alpha' \supset^{+} \beta'| = |T^{S}|$. Hence, the left-hand side of (**) is $\varphi(|T^{S}|)$ is 1, and we are done.

2. $\eta_{T}(\alpha) > \eta_{T}(\beta)$. Then, the right-hand side of (**) is $1 - \frac{\eta_{T}(\alpha)}{s} + \frac{\eta_{T}(\beta)}{s}$, by the definition of φ. According to the definitions of $\to^{*}$ и $\supset^{+}$, the number of T's in $\alpha' \supset^{+} \beta'$ is equal to $\eta_{T}(\beta) + (s - \eta_{T}(\alpha))$. Hence, the left-hand side of (**) is $1 - \frac{\eta_{T}(\alpha)}{s} + \frac{\eta_{T}(\beta)}{s}$, and we are done.

Thus, certain subsets of *T-F*-sequences from B^{S} rather than *T-F*-sequences as such are used as truth-values for Łukasiewicz *n*-valued logic $\mathbf{L_{n}}$. For example, $\{<T,F,F>, <F,T,F>, <F,F,T>\}$ rather than $<T,F,F>$ or $<F,F,T>$ is used to interpret the truth-value 1/3. It is a very interesting by-product of factor-semantics that we hit upon the concept of the structuralization of truth-values.

Incidentally, the cardinality of the set of $|\alpha|$, $|\alpha| \in B^{S}/\cong$, is calculated by formula for binomial coefficients:

$$C^{k}_{m} = \frac{m!}{k!(m-k)!}.$$

In our case, $k = \eta_{T}(\alpha)$ and $m = s$.

Despite considerations in this chapter, we believe that we are able to grasp the essence of Łukasiewicz *n*-valued logics only when $\mathbf{L_{n}}$ are presented as functional systems. From this point of vantage, there are no truth-values, tautologies, and algebraic identities; we only deal with the set of primary (or initial) n-valued functions and the operation of superposition which defined on such a set.

IV. Functional properties of Łukasiewicz n-valued logic

IV.1. Preliminary remarks

In the literature, one can find two basic methods of studying logical systems: *external* one and *internal* one.

With the former, one either (*i*) represents a logical system as a calculus (Hilbert-style, Gentzen-style, etc.) and investigates different semantics of the calculus, or else (*ii*) represents a logical system as an algebra and explores its algebraic properties. In the last analysis, these two approaches converge, forming two sides of the same coin (see, for example, [Blok and Pigozzi, 1989] and [Font, Jansana and Pigozzi, 2003]).

The latter, internal, method is, however, the one which enables us to grasp the very core of $Ł_n$ – internally, we study $Ł_n$ as a functional system. Indeed, by following this latter approach, we will arrive at the connection between functional properties of Łukasiewicz *n*-valued logics $Ł_n$ and prime numbers.

IV.2. J_i-functions

Let's consider the following functions, called J_i-functions, introduced in [Rosser and Turquette, 1952]:

$$J_i(x) = \begin{cases} 1, & \text{if } x = i \\ 0, & \text{if } x \neq i. \end{cases}$$

These functions come in useful in various fields of many-valued logic. Not every many-valued logic, however, enjoys the following remarkable property:

Theorem 1. *J_i-functions are definable through* ~ *and* → *in* $Ł_n$.

For the proof, see [Rosser and Turquette, 1952, pp. 18-22].

J_i-functions are very significant for axiomatization of large classes of many-valued logics, especially see [Anshakov and Rychkov, 1984]. Intuitively, the expressibility in some logic of all J_i-functions means that for every $i \in V_n$, it is possible to say in the language of the logic that a proposition A assumes the given truth-value i.

IV.3.   McNaughton's criterion

It is worth noting that Theorem 1 (as well as numerous other results for both $Ł_n$ and $Ł_\infty$) can be easily obtained as a consequence of an important property of Łukasiewicz logics, *McNaughton's criterion of definability of functions in Łukasiewicz matrices* ([McNaughton, 1951]), which holds for both $Ł_\infty$ and $Ł_n$.

For the finite case, McNaughton proved a theorem stating that, given any natural n and any function f in $\mathfrak{M}_n^L$, we can decide whether f is definable through $\sim x$ and $x \to y$ alone. The theorem boils down to the following: *f is definable in $\mathfrak{M}_n^L$ iff for all $x_1, ..., x_s, x$, if $f(x_1, ..., x_s) = x$, then the greatest common divisor (GCD) of the sequence of numbers $(x_1, ..., x_s, n\text{-}1)$ divides x.*

It should be noted that, even though [McNaughton, 1951] gives a necessary and sufficient condition for a finite-valued function to be definable in $Ł_n$, the proof it provides is not constructive; it only tells which finite-valued functions f are Łukasiewicz-definable, without giving a method of constructing an $Ł_n$-formula defining f in terms of $\sim$ and $\to$. Therefore, it is worth consulting [Takagi, Nakashima, and Mukaidono, 1999], where another necessary and sufficient condition is given; moreover, the latter work explains how to express a function in the language of $Ł_n$ once its truth table is known.[12]

IV.4.   Sheffer stroke for $Ł_n$

Traditionally, special interest attaches to logical systems containing only one truth-function (or, connective).

Following Quine's lead, H.E. Hendry and G.J. Massey ([Hendry and Massey, 1969]) proposed to call a function f an *indigenous* Sheffer stroke for a set of functions F if f is a Sheffer stroke for F (that is, all functions in F are definable through f alone) and, furthermore, f is itself definable through a finite composition of functions in F. Here, we are interested only in *indigenous* Sheffer strokes (see above section I.3.1). I. Rosenberg gave a complete characterization of Sheffer functions in n-valued logics in [Rosenberg, 1978]. It is worth noting that not every set of functions of an n-valued logic has an *indigenous* Sheffer stroke (see [Rose, 1969]); the following theorem, however, can be proved [McKinsey, 1936].

---

[12] It is worth noting that the constructive proof of McNaughton's criterion for $Ł_\infty$ was given in [Mundici, 1994].

Theorem 2. *The function* $Exy = CxC[CNy]yNCyN[Cy]Ny$ *is a Sheffer stroke for* $Ł_n$, *where* C *and* N *are implication and negation in Łukasiewicz notation, and the bracketed parentheses stand for the needed number (i.e. n-2) of occurrences of the embraced expression.*

Let's denote the function Exy as $x \rightarrow^E y$. With the help of functions $J_i(x)$, we can significantly simplify the definition of Exy. To that end, let's note that $J_{n-1}(y) = N\{Cy\}Ny$ and $J_0(y) = N\{CNy\}y$. Then

$$x \rightarrow^E y = x \rightarrow (\sim J_0(y) \rightarrow \sim(y \rightarrow J_1(y))),$$

and, by applying contraposition to the consequent, we get the following:

$$x \rightarrow^E y = x \rightarrow ((y \rightarrow J_1(y)) \rightarrow J_0(y)).$$

For the sake of comparing $x \rightarrow^E y$ with the Łukasiewicz implication $x \rightarrow y$, let's also note that $x \rightarrow^E y$ can also be defined thus:

$$x \rightarrow^E y = \begin{cases} 1, & \text{if } y = 0 \\ \sim x, & \text{if } y = 1 \\ x \rightarrow y & \text{otherwise.} \end{cases}$$

McKinsey's definitions (in our notation) of $\sim x$ and $x \rightarrow y$ are as follows:

$(a)\ 1 = (x \rightarrow^E x) \rightarrow^E ((x \rightarrow^E x) \rightarrow^E (x \rightarrow^E x)),$

$(b)\ \sim x = x \rightarrow^E 1,$

$(c)\ x \rightarrow y = x \rightarrow^E (1 \rightarrow^E y).$

Note that Rose in [Rose, 1968] constructed a commutative Sheffer stroke for $Ł_n$.

IV.5. Functional extensions of $Ł_n$

The following generalization of Słupecki's function Tx (see section I.3.1) was introduced in [Rosser and Turqutte, 1952, pp. 23-25]:

Let $T_{\frac{n-2}{n-1}}(x) = \frac{n-2}{n-1}$ *be for all* $x \in V$. *Then, the system of functions* $\{x \rightarrow y, \sim x, T_{\frac{n-2}{n-1}}(x)\}$ *is functionally complete.*

A generalization of this result gives us the following theorem ([Evans and Schwartz, 1958]):

Let $T_{\frac{i}{n-1}}(x) = \frac{i}{n-1}$, *for* $0 < i < n-1$. *Then, the system of functions* $\{x{\to}y,$
$\sim x,$ $T_{\frac{i}{n-1}}(x)\}$ *is functionally complete iff* $(n-1, i) = 1$, *i.e.* $n-1$ *and* i *are relatively prime numbers.*

The notion of functional completeness is defined in the next section.

IV.6.　　Post logics P_n

Post logics P_n [Post, 1921] are finite-valued logics that are, unlike $Ł_n$, functionally-complete. Post was inspired to devise P_n by the well-known formalization of the classical logic as presented in *Principia Mathematica* by A.N. Whitehead and B. Russell, where only negation ($\neg$) and disjunction ($\vee$) are taken as primitive connectives. The primary objective of Post was a generalization for an arbitrary finite set of truth-values of the classical propositional logic as described in *Principia Mathematica*.

The standard definition of Post's n-valued matrix logics looks as follows. A matrix of the form

$$\mathfrak{M}_n^P = <V_n, \neg, \vee, \{n\text{-}1\} >$$

is called a *Post n-valued matrix* $(n \in N, n \geq 2)$ provided that

$$V_n = \{0, 1, 2, ..., n\text{-}1\};$$

$\neg$ (negation) is a unary function and $\vee$ (disjunction) is a binary function defined on V_n as follows:

$$\neg x = x + 1(mod\ n),$$
$$x \vee y = max(x, y).$$

and $\{n\text{-}1\}$ is the set of the designated elements of $\mathfrak{M}_n^P$.

$\neg x$ is usually called a cyclical negation; its truth-table looks thus:

x	$\neg x$
0	1
1	2
.	.
.	.
.	.
n-2	n-1
n-1	0

It is easily seen that the two-valued Post matrix is isomorphic to the negation and disjunction matrices of the classical propositional logic C_2 (in exactly the same way as the two-valued Łukasiewicz matrix is isomorphic to the negation and implication matrices for C_2). The matrices for any $\mathfrak{M}_n^P$ are, however, in a sense, totally incompatible with C_2-matrices, due to the non-standard way $\neg$ is defined in $\mathfrak{M}_n^P$. Thus, for $n = 3$ we have that the formula

$$p \vee \neg p \vee \neg\neg p$$

is a tautology.

Generally, P_n verifies 'the generalized law of excluded middle'.

Remember that classical propositional logic C_2 is functionally complete. We have already seen in the previous chapter that $Ł_n$ are not functionally complete. In this respect, P_n is like C_2.

Theorem 3. *Post's n-valued logic P_n is functionally complete, i.e. each k-argument operation on the set of truth-values of P_n can be defined using the operations $\neg$ and $\vee$* [Post, 1921][13].

In [Webb, 1936], the Webb's stroke operation is defined through primitive operations of P_n in the following way:

$$W_n(x, y) = \neg(x \vee y), \text{ or}$$

$$W_n(x, y) = max(x, y) + 1(mod\, n).$$

Hilbert-style axiomatization of P_n can be found in H. Rasiowa [Rasiowa, 1974, ch. XIV].

---

[13] Two more proofs of this result can be found in [Barton, 1979].

IV.7.   Logic as a functional system:

Closure operator, completeness and precompleteness

A function $f(x_1, ..., x_s)$ with a finite number of arguments is called an $n+1$-valued function, or a function of $n+1$-valued logic, if f is a map from the power set V_{n+1}^s into V_{n+1}, where $V_{n+1} = \{0, 1, 2, ..., n\}$. Let P_{n+1} be the set of all $n+1$valued functions defined on the set V_{n+1}. Then, a pair (P_{n+1}, C), where C is the operation of *superposition* of functions, is a functional system. Roughly speaking, the result of superposition of functions $f_1, ..., f_k$ is the function obtained from $f_1, ..., f_k$ either (1) by substituting some of these functions for arguments of $f_1, ..., f_k$ or (2) by renaming arguments of $f_1, ..., f_k$ or by both (1) and (2). An example of a functional system is $(Ł_{n+1}, C)$, where $Ł_{n+1}$ is the set of all $(n+1$-valued) functions of Łukasiewicz logic $\mathbf{L}_{n+1}$.

We will employ the following terminology, introduced by A.V. Kuznetsov (see [Janowskaja, 1959]). Let $F \subseteq P_{n+1}$. We define a closure operator [ ] on the power-set of P_{n+1}, in the following way (intuitively, $[F]$ is the set of all superpositions of functions from F):

(*i*) $F \subseteq [F]$,

(*ii*) $[[F]] = [F]$,

(*iii*) if $F_1 \subseteq F_2$ then $[F_1] \subseteq [F_2]$.

A set F of functions is said to be closed if $F = [F]$. A set F of functions is (functionally) complete in $\mathfrak{R} \subseteq P_{n+1}$ (where $\mathfrak{R}$ is a closed set of functions), if $[F] = \mathfrak{R}$. Lastly, a closed set F is called *precomplete* in P_{n+1}, if $[F] \neq P_{n+1}$ and $[F \cup \{f\}] = P_{n+1}$, where $f \in P_{n+1}$ and $f \notin F$ (in other teminology, a precomplete class of functions is called *maximal clone*).

As an example, let T_{n+1} ($n \geq 2$) be the set of all functions from P_{n+1} which preserve 0 and n, i.e. $f(x_1, ..., x_s) \in T_{n+1}$ iff $f(x_1, ..., x_s) \in \{0, n)$, where $x_i \in \{0, n\}$, $0 \leq i \leq s$. S. Jablonski in [Jablonski, 1958] showed that T_{n+1} is precomplete in P_{n+1} for each $n \geq 2$. Precomplete classes of functions are vitally important for the characterization of functional completeness: *a set F of n+1-valued functions is functionally complete iff it is not contained in a precomplete set of functions*. A complete characterization of precomplete classes of functions has been provided by I. Rosenberg in 1965 (see [Rosenberg, 1970]).

IV.8. $Ł_3$ between C_2 and P_3: the continuality of $Ł_3$

A paramount task in studying multi-valued logics as functional systems is a description of the lattice of the closed classes of a given logic. For the classical logic this problem was completely solved by E. Post in the 1920s. In [Post, 1921], it was proved that the set of closed classes of P_2 (or *all possible clones on {0,1}*) is countable, while in [Post, 1941] a complete description of the lattice of closed classes is given in such a way that every closed class is effectively constructed and each class has a finite base. These classes are usually referred to as *Post classes*. The classification of E. Post was presented in a more modern notation by R. Lyndon [Lyndon, 1951]. In [Post, 1941], the question is also raised concerning a description of the closed classes of P_n.

Many-valued logic, however, proved to be quite different from the classical one; these differences indicate that the former *can not be reduced* to the latter. For example, it follows from the work of Y. I. Janov and A. A. Muchnik [Yanov and Muchnik, 1958] (see also [Hulanicki and Swierckowski, 1960]) that, *for every $n \geq 3$, P_n has a continuum of distinct closed classes*, that is even P_3 already has a continuum of closed classes (clones on {0, $^1/_2$,1}. The exact cause of such a difference between two- and multi-valued logics seems to be unclear.

Since P_3 has a continuum of closed classes, it is interesting to ask what cardinality is the set of closed classes of other three-valued logics. Of special interest, because of her close connection to the intuitionistic propositional logic **H**, in this respect is the three-valued Heyting's logic G_3 (1930), also called Jaśkowski's first matrix [Jaśkowski (1936), 1967]:

x	$\rceil x$
1	0
$^1/_2$	0
0	1

$\Rightarrow$	1	$^1/_2$	0
1	1	$^1/_2$	0
$^1/_2$	1	1	0
0	1	1	1

Thus matrices for $\vee$ and $\wedge$ are exactly the same as the matrices defining these connectives in $Ł_3$, that is

$$x \vee y = max(x, y)$$
$$x \wedge y = min(x, y).$$

It is clear that H_3 is different from $Ł_3$ since $\sim x$ is not definable in G_3. Thus, $G_3 \subset Ł_3$. If, however, we add the function $\sim x$ to G_3, then we obtain $Ł_3$:

$$x \rightarrow y = (x \Rightarrow y) \vee \sim x) \quad \text{[Cignoli, 1982, p. 9]}.$$

Let's notice the following result obtained by M. Ratsa [Ratsa, 1982], which is a consequence of his more general results: G_3 has a *continuum* of distinct closed classes.

As G_3 is functionally embeddable into $Ł_3$,:

$$\daleth x = \sim(\sim x \rightarrow x),$$

$$x \Rightarrow y = \daleth (\sim(x \rightarrow y)) \vee y,$$

the same is true of $Ł_3$. Moreover, Ratsa showed that G_3 is a pre-complete class in $Ł_3$.

It thus turned out that, to pull down the doctrine of logical fatalism, Łukasiewicz abandoned discreteness for continuity. As a result, we have to deal with continuality of Łukasiewicz logics $Ł_n$.

IV.9. Maximal n+1-valued non-Postian logic

V. K. Finn in [Finn, 1975] formulated, in the form of a logical matrix, a logic **T_{n+1}**, which was later called ([Bocvar and Finn, 1976, p. 266]) a *"maximal n+1-valued non-Postian logic"*. **T_{n+1}** is defined as follows:

$$\mathfrak{M}^{T}_{n+1} = \langle V_{n+1}, \sim x, x \wedge y, x \vee y, J_0(x), ..., J_n(x), N_1(x), ..., N_{n-1}(x), \{n\}\rangle, \text{ where}$$

$\sim x$, $x \wedge y$, and $x \vee y$ are defined as in Łukasiewicz logics, and $J_i(x)$ are the functions defined in section II.1; $N_i(x)$ are defined thus:

$$N_i(x) = \begin{cases} i, \text{ if } x \in \{1,...,n-1\} \\ \sim x, \text{ if } x \in \{0,n\} \end{cases} \quad (1 \leq i \leq n-1).$$

We can appreciably simplify the signature of $\mathfrak{M}^{T}_{n+1}$. Let $\mathfrak{M}^{T^*}_{n+1}$ be

$\langle V_{n+1}, \sim x, x \rightarrow^{T^*} y, \{n\}\rangle$, where

1) if $n = 2$, then $x \rightarrow^{T^*} y = x \rightarrow y$;

2) if $n > 2$, then

$$x \rightarrow^{T^*} y = \begin{cases} n-1, \text{ if } x = y \text{ and } x, y \in \{1,...,n-1\} \\ x \rightarrow y, \text{ otherwise.} \end{cases}$$

By T^*_{n+1} we denote the set of functions of $\mathfrak{M}^{T^*}_{n+1}$.

Theorem 4. $T^*_{n+1} = T_{n+1}$ *for any* $n \geq 2$.

There follows a more detailed proof of Theorem 4 than the one given in [Karpenko, 1983], where it was shown that the matrix logic T^*_{n+1} has a factor-semantics.

I. First, we show that $T_{n+1} \subseteq T^*_{n+1}$.

Let's begin by defining Łukasiewicz implication $x \to y$ in T^*_{n+1}:

$$x \to y = {\sim}((y \to^{T^*} x) \to^{T^*} {\sim}(y \to^{T^*} x)) \to^{T^*} (x \to^{T^*} y).$$

It is easily seen that

$$x \to y = {\sim}((y \to x) \to {\sim}(y \to x)) \to (x \to y).$$

Since $x \to^{T^*} y$ differs from $x \to y$ only when $x = y$ and $x, y \in \{1, \ldots, n\text{-}1\}$, all we have to prove is the following:

$$x \to y = {\sim}((n\text{-}1) \to^{T^*} ((n)\text{-}(n\text{-}1))) \to^{T^*} (n\text{-}1) =$$
$$= {\sim}((n\text{-}1) \to^{T^*} 1) \to^{T^*} (n\text{-}1) = ((n)\text{-}1) \to^{T^*} (n\text{-}1) = n.$$

It follows that $Ł_{n+1} \subseteq T^*_{n+1}$. Since $x \wedge y \in Ł_{n+1}$, we have $x \wedge y \in T^*_{n+1}$. In virtue of the theorem by Rosser and Turquette (section II.1), we also have $J_i(x) \in Ł_{n+1}$. Therefore, $J_i(x) \in T^*_{n+1}$. What remains to be shown is that $N_i(x) \in T^*_{n+1}$. There are two cases to consider.

1) $n = 2$. Then

$$N_1(x) = {\sim}x;$$

2) $n > 2$. Then

$$N_1(x) = (x \to^{T^*} x) \to^{T^*} J_0(x),$$
$$N_2(x) = (x \to^{T^*} x) \to^{T^*} N_1(x),$$

. . . .

$$N_{n\text{-}1}(x) = (x \to^{T^*} x) \to^{T^*} N_{n\text{-}2}(x),$$

Thus, $T_{n+1} \subseteq T^*_{n+1}$.

II. Secondly, we prove that $T^*_{n+1} \subseteq T_{n+1}$.

We have already shown that T^*_{n+1} includes T_{n+1}. But T_{n+1} is functionally precomplete in P_{n+1} for any $n \geq 2$. Since T^*_{n+1} is not functionally complete in P_{n+1} (the functions $\sim x$ and $x \to^{T^*} y$ preserve the set of values $\{0, n\}$), we have that $T^*_{n+1} \subseteq T_{n+1}$.

This completes the proof.

In conclusion, we note that in [Karpenko, 1989, p.180] a Sheffer stroke for T^*_{n+1} is defined.

IV.10. Precompleteness and prime numbers

V.K. Finn in [Finn, 1969] showed that the set of functions of $Ł_3$ is functionally precomplete in P_3, i.e. $Ł_3 = T_3$.[14] Therefore, there arises the question of the criterion of functional precompleteness for the set of functions $Ł_{n+1}$ for arbitrary n – i.e. the question as to for which n we have $Ł_{n+1} = T_{n+1}$. The solution was given by Finn in the short note [Finn, 1970], and the problem was thoroughly investigated in [Bochvar and Finn, 1972] (an English abstract for this work has been published as [Finn, 1975]).

Let $I_{\xi\eta}(x)$ be a set of functions defined as follows:

$$I_{\xi\eta}(x) = \begin{cases} \eta, & if\ x = \xi \\ 0, & if\ x \neq \xi \end{cases} \quad (0 < \xi, \eta < n).$$

Truth-tables for these functions look like this:

x	0	1	...	i	...	n-1	n
$I_{\xi\eta}(x)$	0	0	...	j	...	0	0,

where $\xi = i$, $\eta = j$, $1 \leq i, j \leq n$-1.

Let I_{n+1} be the set of all $I_{\xi\eta}(x)$-functions definable in T_{n+1}.

Theorem 5. *The set of functions $Ł_{n+1}$ is functionally precomplete in P_{n+1} if and only if $I_{\xi\eta}(x) \subset Ł_{n+1}$; then, $Ł_{n+1} = T_{n+1}$* [Bochvar and Finn, 1972, pp. 248-253].

---

[14] This result was rediscovered at least twice (see [Hersberger, 1977] and [Hendry, 1980]).

The proof of Theorem 5 consists of the proof of the following statement: any function $f \in T_{n+1}$ that is not equal to the constant function 0, is definable by a superposition of $x \vee y = max(x,y)$, $x \wedge y = min(x,y)$, I-functions and J-functions (see Theorem 2). This superposition is an analogue of the full disjunctive normal forms of the two-valued logic (we denote this superposition by I-J-f.d.n.f).

Now, for which n, $I_{n+1} \subset Ł_{n+1}$? Answering this question will answer our initial question as to for which n does the equation $Ł_{n+1} = T_{n+1}$ hold? The answer is supplied by the following theorem.

Theorem 6. (*A criterion of precompleteness for sets of functions $Ł_{n+1}$*) $Ł_{n+1}$ $= T_{n+1}$ *if and only if n is a prime number* [Bochvar and Finn, 1972, pp. 255-276] [15].

Theorem 6 follows from Theorem 5 and McNaughton's criterion (see section II.2). It is worth noting that [Bochvar and Finn, 1972] contains a constructive proof of Theorem 6, a proof that does not directly use McNaughton's result. Obviously, if n is a prime umber, then GCD of numbers $(x_1, \ldots, x_s, n)$ is 1; hence, $T_{n+1} \subset Ł_{n+1}$. Thus, as an example, the connectives of $Ł_{12}$ make up a precomplete set, while those of $Ł_{13}$ do not.

Along these lines, a new definition of a prime number can be given: a natural number $n \geq 2$ is prime iff the set of all functions corresponding to $n+1$-valued Łukasiewicz logic is a precomplete set in P_{n+1}, that is $Ł_{n+1} = T_{n+1}$.

From Theorem 6 and from Tarski's theorem on cardinal degrees of completeness of $Ł_{n+1}$ (see section III.4) there follows

Corollary 1. *If $Ł_{n+1} = T_{n+1}$, where $n \geq 2$, then $\gamma(Ł_{n+1}) = 3$.*

---

[15] This result was rediscovered twice (see [Hendry, 1983] and [Urquhart, 1986, pp. 87-89]). The last proof is a very elegant and directly based on McNaughton's criterion which was also proved by him for this case.

V.   Structuralization of prime numbers

V.1.   Partition of the set of Łukasiewicz logics Ł$_{n+1}$ relative to the precompleteness property

In the context of the preceding chapter, the following question arises: can we build a sequence of $n+1$-valued logics such that all of them whose n is equal or greater than 2 are precomplete, but such that all logics in this sequence have all the properties of Łukasiewicz logics $Ł_{n+1}$? Apparently, this contradicts Theorem 6; nevertheless, in a sense, such a sequence can be constructed; in the present section, we show how.

Among the three proofs of Theorem 6 mentioned in the previous chapter (Finn's, Hendry's, and Urquhart's), only that by Finn shows explicitly that $Ł_{n+1}$ is not precomplete when $n \neq p$, where p is a prime number, since not all $I_{\xi\eta}(x)$-functions are definable in $Ł_{n+1}$ (see section II.8).

For example, let $n + 1 = 10$; for convenience, denote the set V_{10} of truth-values as

$$\{0, \, {}^1/_9, \, {}^2/_9, \, {}^3/_9, \, {}^4/_9, \, {}^5/_9, \, {}^6/_9, \, {}^7/_9, \, {}^8/_9, \, 1\}.$$

In virtue of McNaughton's criterion, it follows that it is impossible to define in $Ł_{10}$ the function $I_{\frac{3}{9}\frac{7}{9}}(x)$ when $x = {}^3/_9$ (then $I_{\frac{3}{9}\frac{7}{9}}({}^3/_9) = {}^7/_9$), since GCD $(3, 9) = 3$ but GCD $(7, 9) = 1$; hence, $Ł_{10}$ is not precomplete. Clearly, if numerators and denominators for all ${}^i/_n$ from V_{n+1} are relatively prime numbers, i.e. GCD $(i, n) = 1$, then the definability of I-functions in $Ł_{n+1}$ is preserved. It follows, then, that the values ${}^3/_9$ and ${}^6/_9$, are responsible for the non-precompleteness of $Ł_{10}$ since the numerators and denominators of these values are not relatively prime numbers. Removing from V_{10} these "bad" values, we are left with just eight truth-values. Let be $n = 8$, then

$$V_8 = \{0, \, {}^1/_7, \, {}^2/_7, \, {}^3/_7, \, {}^4/_7, \, {}^5/_7, \, {}^6/_7, \, 1\}.$$

Thus, we made a transition from $Ł_{10}$ to $Ł_8$, and in virtue of Theorem 6, $Ł_8$ is a precomplete logic. So, to reconstruct some arbitrary set of truth-values V_{n+1}, we have to identify how many numbers i in the set $\{1, 2, \ldots, n-1\}$ are relatively prime to n, and to add two "limit" numbers, 0 and 1.

A function identifying the number of positive integers $\leq n$ that are relatively prime to n (1 is considered relatively prime to all numbers) is called the *totient function* $\varphi(n)$, also called *Euller's totient function* (introduced in 1760)[16].

Examples:

$\varphi(1) = 1,$	$\varphi(5) = 4,$	$\varphi(9) = 6,$
$\varphi(2) = 1,$	$\varphi(6) = 2,$	$\varphi(10) = 4,$
$\varphi(3) = 2,$	$\varphi(7) = 6,$	$\varphi(11) = 10,$
$\varphi(4) = 2,$	$\varphi(8) = 4,$	$\varphi(12) = 4.$

Basic properties of totient function $\varphi(n)$:

(*i*) $\varphi(1) = 1.$

(*ii*) $\varphi(n)$ is multiplicative function, i.e. if $(n_1, n_2) = 1$,

then $\varphi(n_1 \cdot n_2) = \varphi(n_1) \cdot \varphi(n_2)$.

(*iii*) For any prime number p, $\varphi(p^\beta) = p^{\beta-1}(p-1)$.

A general formula follows from (*ii*) and (*iii*):

(*iv*) $\varphi(n) = n(1 - \frac{1}{p_1}) \cdot (1 - \frac{1}{p_2})...(1 - \frac{1}{p_s}) = n\prod_{p|n} (1 - \frac{1}{p}),$

(*v*) $\varphi(n)$ is always even for $n \geq 3$.

(*vi*) If for some n, $\varphi(n) = m$, then $\varphi(2n) = m$ iff n is odd.

(*vii*) $\varphi(p) = p-1$, for prime p.

In the language of group theory, $\varphi(n)$ is the number of generators in a cyclic group of order n.

Not every even number is a value of $\varphi(n)$. For example, numbers

$$14, 26, 34, 38, 50, 62, 68, 74, 76, 86, 90, 94 \text{ and } 98$$

are not values of $\varphi(n)$, when $n \leq 100$. Moreover, from Rechman's result it follows that there are infinitely many even numbers that are not values of $\varphi(n)$ [Rechman, 1977].

Let $\varphi^*(n) = \varphi(n)+1$. If n is some prime number p_s, then $\varphi^*(p_s) = (p_s-1)+1 = p_s$. Hence, if $Ł_{n+1}$ is precomplete, then the application of the function $\varphi^*(n)$ does not affect the precompleteness of $Ł_{n+1}$.

Note that, for the above-mentioned examples of the values of totient function $\varphi(n)$, $\varphi^*(n) = p$, but this does not hold good for all n. For example,

---

[16] See http://mathworld.wolfram.com/TotientFunction.html, which contains an extensive number of references; also see [Maier and Pomerance, 1988] and [Spyroponlus, 1989].

$\varphi^*(16) = 9$, but $Ł_{9+1}$ is not a precomplete logic since $9 \neq p$. However, as we know, $\varphi^*(9) = 7$ and, consequently, $Ł_{7+1}$ is a precomplete logic. Thus, constructing a precomplete logic $Ł_{p+1}$ from an arbitrary logic $Ł_{n+1}$ comes down to the transformation of number n into a prime number p.

Now, let's denote the result of k iterations of $\varphi^*(n)$ by $\varphi_k^*(n)$. Clearly, *for each n there exists k such that* $\varphi_k^*(n)$ *is a prime number p*, i.e.

$$\forall n \exists k (\varphi_k^*(n)) = p.$$

Thus, we have the *algorithm* transforming every logic $Ł_{n+1}$ into a precomplete logic $Ł_{p+1}$, and consequently, transforming every natural number n into a prime number p.

By this algorithm, the logic $Ł_{35+1}$ is transformed into the precomplete logic $Ł_{13+1}$; in this case, $k = 3$.[17] The function $\varphi_k^*(n)$ generates the infinite consequence of prime numbers [Karpenko, 1983, p. 105]:

$$2, 2, 3, 3, 5, 3, 7, 5, 7, 5, 11, 5, 13, 7, 7, 7, 17, 7, 19, 7,$$
$$13, 11, 23, 7, 13, 13, 19, 13, 29, 7, 31, \dots$$

It is worth mentioning here that only in August 2000 the author came across, on the web-site [Sloane, 1999][18], the very same consequence, referred to as "A0399650". A further development of this result will prove very important later on.

The above-mentioned algorithm gives us a partition of the set of logics $Ł_{n+1}$ into equivalence classes so that every class contains one and only one precomplete logic $Ł_{p_s+1}$, i.e.

$$Ł_{n_1+1} \cong Ł_{n_2+1} \text{ iff } \exists k \exists l (\varphi_k^*(n_1) = \varphi_l^*(n_2)).$$

These classes are denoted by X_{p_s+1}. For example, $X_{p_3+1} = \{6, 9, 11, 13\}$, where $p_3 = 5$.

---

[17] For any $n \leq 100000$ we can calculate the value $\varphi_k^*(n)$ using the table of values $\varphi(n)$ given in [Lal and Gillard, 1968].

[18] This web-site is an extended version of the book [Sloane and Plouffe, 1995].

V.2. Construction of the classes $\mathcal{X}_{p_s+1}$ (inverse Euler's totient function)

Linked with the above-mentioned partitioning of Łukasiewicz logics $Ł_{n+1}$ is the problem of constructing an equivalence class $\mathcal{X}_{p+1}$ for an arbitrary precomplete logic $Ł_{p+1}$. To solve this problem, we need a function inverse to $\varphi_k^*(n)$. To this end, we have to know the set of values of the *inverse totient function* $\varphi^{-1}(m)$, defined by

$$\varphi^{-1}(m) = \{n: \varphi(n) = m\}.$$

Thus, if $\varphi(n) = 4$, the equation has exactly four solutions, i.e. the set of values of $\varphi^{-1}(4)$ is $\{5, 8, 10, 12\}$.

Probably, the first to pay attention to the problem of solving such equation was E. Lucas (1842-1891). The table of values of $\varphi^{-1}(m)$ for $m \leq 2500$ was published in [Glaisher, 1940], as late as in 1940s. Our table 2 extends that table up to $m \leq 5000$.

It is interesting to note that in [Bolker, 1970] the exercise (No. 11.19) was suggested of finding all solutions of the equation $\varphi(n) = 24$. The solution was given in [Burton, 1976, p. 350]:

$$\varphi^{-1}(24) = \{35, 39, 45, 52, 56, 70, 72, 84, 90\}.$$

The first work especially devoted to the properties of inverse Euler's totient function appeared as late as in 1981 [Gupta, 1981].[19] This function was referred to in that work as $\varphi^{-1}(m)$.

The set $\varphi^{-1}(m) = \{n: \varphi(n) = m\}$ is empty for all odd values of $m > 1$ and also for many even values of m. Here, we are only interested in those values of $\varphi^{-1}(m)$ that are non-empty. Obviously, any such set is finite since the number of divisors of m is finite. In [Gupta, 1981] the following theorem was proved:

Theorem 1. *Any non-empty set $\varphi^{-1}(m)$ is bounded both above and below.*

Most important for us, H. Gupta described a method of identifying all elements of the set $\varphi^{-1}(m) = \{n: \varphi(n) = m\}$. Let n be an element of $\varphi^{-1}(m)$ for some given m. Assume that p is the least prime divisor of n. Let

---

[19] It is worth noting that in that work a simple reduction formula for $\varphi(n)$ is given: $\varphi(n) = p\varphi(u)$ or $(p-1)\varphi(u)$ according to whether on not p divides u. With $\varphi(1) = 1$, this formula completely defines $\varphi(n)$ for all positive integral values of n. This formula had never appeared in print before Gupta's paper.

$$n = p^d u, \text{ where } (u, p) = 1.$$

This clearly implies that u has no prime divisor $\leq p$. Evidently, we have

$$(1) \quad m = \varphi(n) = \varphi(p^d)\, \varphi(u).$$

For (1) to hold, p has to be such that

$$(2) \quad (p-1) \big| m$$

and u should belong to the subset of $\varphi^{-1}(m)/\varphi(p^d))$ containing those elements that have no prime divisor $\leq p$. Such a subset we denote by $\varphi_p^{-1}(m/\varphi(p^d))$. It will be clear that every element of

$$(3) \quad p^d \varphi_p^{-1}(m/\varphi(p^d))$$

gives a solution of the equation

$$(4) \quad \varphi(x) = m.$$

In fact, (3) provides all solutions of (4) that have p as their least prime divisor and p^d as the highest power of p that divides them.

By letting p range over all the primes satisfying condition (2) and d over all the values for which $\varphi(p^d)$ divides m, all solutions of (4) can be obtained. These determine $\varphi^{-1}(m)$. For any prime p satisfying (2), we can ignore all the values of d for which $m/\varphi(p^d)$ is an odd number > 1.

For reasons which will become clear later on, it is most convenient to consider the values of p in a descending, and those of d in an ascending, order.

The following example may clarify the procedure.

Example. Take $m = 36$ (Gupta takes $m = 576$).

To get the primes p for which $(p-1)\big| m$, we write out all divisors of m; add 1 to each of them and, lastly, take only those of the resultant numbers that are primes. Now, $36 = 2^2 \cdot 3^2$, the divisors of 36, therefore are:

$$1, 2, 4; 3, 6, 12; 9, 18, 36.$$

Adding 1 to each of these, we get

$$2, 3, 5; 4, 7, 13; 10, 19, 37.$$

The primes among these, arranged in descending order, are:

$$37, 19, 13, 7, 5, 3, 2.$$

We assume that sets $\varphi^{-1}(x)$ are available for all $x < 36$. Those that we need are:

x	$\varphi^{-1}(x)$
1	$\{1, 2\}$
2	$\{4, 6\}$
6	$\{7, 9, 14, 18\}$
18	$\{19, 27, 38, 54\}$

Our calculations can now be presented in the following tabular form:

p	d	$m/\varphi(p^d)$	$p^d \varphi_p^{-1}(m/\varphi(p^d))$
37	1	1	$37\{1\} = \{37\}$
19	1	2	$19.\varnothing$
13	1	3	$-$
7	1	6	$7.\varnothing$
5	1	9	$-$
3	1	18	$3\{19\} = \{57\}$
3	2	6	$9\{7\} = \{63\}$

At the next stage, we need all the odd elements of $\varphi^{-1}(36)$ and these are already available. (This explains why we consider the primes in descending order.)

p	d	$m/\varphi(p^d)$	$p^d \varphi_p^{-1}(m/\varphi(p^d))$
2	1	36	$2\{37, 57, 63\} = \{74, 114, 126\}$
2	2	18	$4\{19, 27\} = \{76, 108\}$
2	3	9	$-$

Finally, $\varphi^{-1}(36) = \{37, 57, 63, 74, 76, 108, 114, 126\}$.

It is worth noting that, in our calculations, the even elements of the sets mentioned earlier do not play any role.

Let's take m that is not too small. In such cases, as Gupta established, there are twice as many even as there are odd solutions of $\varphi(x) \leq m$.

Admittedly, the described method of calculating the set $\varphi^{-1}(m)$ is rather cumbersome; all the more so given that to find the set $\varphi^{-1}(m)$ we have to know $\varphi^{-1}(x)$ for all $x < m$ and to have the prime factorization of the elements from $\varphi^{-1}(x)$. Nevertheless, the method *is* effective.

Some heuristics in applying this method can be suggested. Take, again, $m = 36$. Write out all divisors of 36: 1, 2, 3, 4, 6, 9, 18, 36. Let's consider different representations of 36 by the products of these divisors, *but those divisors that are values of* $\varphi(n)$; for example, $2 \cdot 18 = 36$. Since $\varphi(3) = 2$, $\varphi(19) = 18$, and $(3, 19) = 1$, we have that $\varphi(3) \cdot (19) = \varphi(3 \cdot 19) = \varphi(57)$ (see *ii*). Hence, $n = 57$. As n is odd, the value of $\varphi^{-1}(36)$ will be also $2 \cdot n = 114$ (see *vi*).

Finally, we have:

$m = 36 = \varphi(37), n = 37;$

$m = 36 = 1 \cdot 36 = \varphi(2) \cdot \varphi(37) = \varphi(2 \cdot 37) = \varphi(74), n = 74;$

$m = 36 = 2 \cdot 18 = \varphi(3) \cdot \varphi(19) = \varphi(3 \cdot 19) = \varphi(57), n = 57;$

$m = 36 = 1 \cdot 2 \cdot 18 = \varphi(2) \cdot \varphi(3) \cdot \varphi(19) = \varphi(2 \cdot 3 \cdot 19) = \varphi(114),$
    $n = 114;$

$m = 36 = 2 \cdot 1 \cdot 18 = \varphi(2^2) \cdot \varphi(19) = \varphi(2^2 \cdot 19) = \varphi(76), n = 76;$

$m = 36 = 2 \cdot 1 \cdot 3^2 \cdot 2 = \varphi(2^2) \cdot \varphi(3^3) = \varphi(2^2 \cdot 3^3) = \varphi(108), n = 108;$

$m = 36 = 3 \cdot 2 \cdot 6 = \varphi(3^2) \cdot \varphi(7) = \varphi(3^2 \cdot 7) = \varphi(63), n = 63;$

$m = 36 = 1 \cdot 3 \cdot 2 \cdot 6 = \varphi(2) \cdot \varphi(3^2) \cdot \varphi(7) = \varphi(2 \cdot 3^2 \cdot 7) = \varphi(126), n = 126.$

The existence of the effective method of calculating the set $\varphi^{-1}(m)$ makes it possible, at least in principle, to construct an algorithm which builds, given a prime number p, its equivalence class X_p (or, given a precomplete logic $Ł_{p+1}$, its equivalence class X_{p+1}). The basic idea of the algorithm is as follows (see [Karpenko, 1983, p. 106]). Consider the function $\varphi_l^{*-1}(m)$, the inverse of $\varphi_k^*(n)$. Let m be some prime p.

1. $\varphi_l^{*-1}(p) = \{v_e\}_1 \cup \{v_o\}_1$ since $\varphi^{-1}(p - 1) = \{v_e\}_1 \cup \{v_o\}_1$, where $\{v_e\}_1$ is the set of even values, and $\{v_o\}_1$ is the set of odd values,

except p. The values in $\{v_e\}_i$ are at every stage of the algorithm discarded, since v_e-1, being an odd number, can not be a value of Euler's totient function $\varphi(n)$. If $\{v_0-1\}_1$ are not values of $\varphi(n)$, then $\varphi_2^{*-1}(v_0)_1 = \varnothing$. Hence, an equivalence class X_p has been constructed; otherwise, go to

2. $\quad \varphi_2^{*-1}(v_0)_1 = \{v_e\}_2 \cup \{v_0\}_2.$

.
.
.

k. $\quad \varphi_k^{*-1}(v_0)_{k-1} = \{v_e\}_k \cup \{v_0\}_k.$ If $\varphi_{k+1}^{*-1}(v_0)_k \neq \varnothing$, then

.
.
.

l, $\quad$ where $k \leq l$.

Example. Let p be 37.

As was established earlier on, $\varphi^{-1}(36) = \{37, 57, 63, 74, 76, 108, 114, 126\}$, i.e. $\varphi_1^{*-1}(37) = \{74, 76, 108, 114, 126\} \cup \{57, 63\}$, where $\{v_e\}_1 = \{74, 76, 108, 114, 126\}$ and $\{v_0\}_1 = \{57, 63\}$. Let's consider $\varphi_2^{*-1}(57)$ and $\varphi_2^{*-1}(63)$. As the set of values of $\varphi_2^{*-1}(63)$ is empty since the number 62 is not a value of $\varphi(n)$, only $\varphi_2^{*-1}(57)$ remains to be considered. Now, $\varphi_2^{*-1}(57) = \{116, 174\} \cup \{87\}$, where $\{v_e\}_2 = \{116, 174\}$ and $\{v_0\}_2 = \{87\}$. Since $\varphi_3^{*-1}(87) = \varnothing$, the construction of the class X_{37} is completed.

V.3. Graphs for prime numbers

With the help of the algorithm building, for a prime number p, its equivalence class X_p, we can represent every class X_p in the form of a *rooted tree*, i.e. a connected acyclic graph, denoted as $\mathcal{T}_p$, with a single special vertex (the root); the vertices of the graph are elements of X_p, and the root is the prime member of X_p.

Examples. The graphs for the first five prime numbers are:

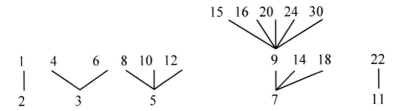

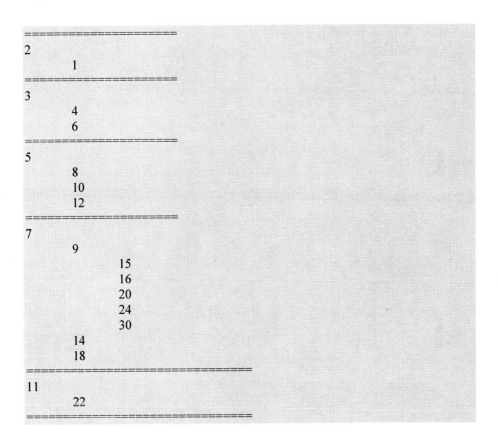

Figure 1.

Further on, we will represent such rooted trees as it is usually done on the screen of a computer terminal; thus, the above five graphs are represented thus:

```
================================
2
     1
================================
3
     4
     6
================================
5
     8
     10
     12
================================
7
     9
               15
               16
               20
               24
               30
          14
          18
================================
11
     22
================================
```

In [Karpenko, 1983, p. 107] the rooted trees for the first thirteen prime numbers were given. Here, we extend them up to the first twenty-five prime numbers (these are exactly prime numbers of the first hundred of the set of natural numbers):

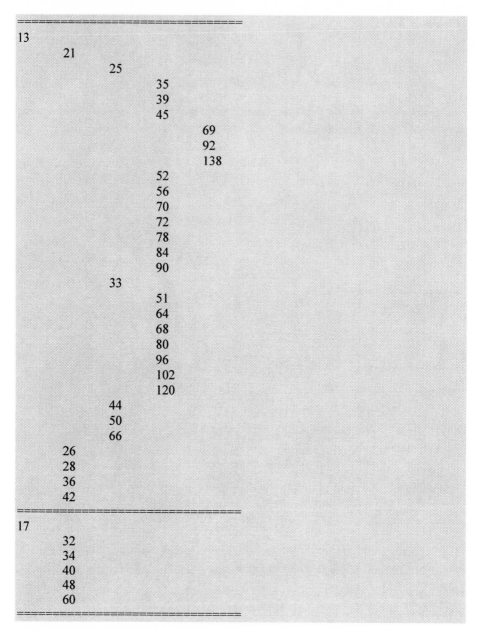

======================================
19
    27
    38
    54
======================================
23
    46
======================================
29
    58
======================================
31
    62
======================================
37
    57
        87
        116
        174
    63
    74
    76
    108
    114
    126
======================================
41
    55
        81
            123
            164
            165
                249
                332
                498
            176
            200
            220
            246
            264
            300
            330
        162
    75
    82
    88

```
        100
        110
        132
        150
============================
43
      49
            65
                  85
                        129
                              255
                              256
                              272
                              320
                              340
                              384
                              408
                              480
                              510
                        147
                        172
                        196
                        258
                        294
                  128
                  136
                  160
                  170
                  192
                  204
                  240
            104
            105
                  159
                  212
                  318
            112
            130
            140
            144
            156
            168
            180
            210
      86
      98
============================
```

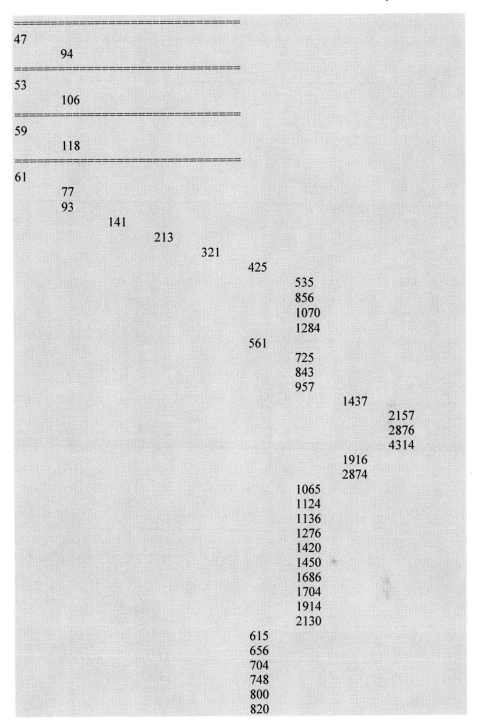

				850
				880
				984
				1056
				1122
				1200
				1230
				1320
			428	
			642	
		284		
		426		
	188			
	282			
99				
122				
124				
154				
186				
198				

=====================================

67
   134

=====================================

71
   142

=====================================

73

91			
95			
111			
	121		
		143	
		155	
		175	
		183	
		225	
			339
			435
			452
			464
			580
			678
			696
			870
		231	
		244	

248
286
308
310
350
366
372
396
450
462
242
117
177
267
345
519
692
1038
356
368
460
534
552
690
236
354
135
146
148
152
182
190
216
222
228
234
252
270
===
79
158
===
83
166
===
89
115

178	
184	
230	
276	

97	
119	
153	
194	
195	
208	
224	
238	
260	
280	
288	
306	
312	
336	
360	
390	
420	

By 1983, using the table of values $\varphi^{-1}(m)$ for $m \leq 2500$ given [Glaisher, 1940], the rooted trees for the first forty-two prime numbers were constructed. By 1989, using our heuristics for Gupta's method the rooted trees for the first fifty prime numbers were built (see [Karpenko, 1989]). Later still, with the help of a computer program, the tree for number 241 (No. 53) was constructed; it looks thus:

241
   287
   305
   325
      489
      513
         771
         1024
         1028
         1088
         1280
         1360
         1536

1542
1632
1920
2040
567
652
972
978
1026
1134
369
705
1059
1173
1761
2057
2225
2785
3029
3835
4893
7341
11013
16521
20675
27291
36388
41350
54582
22028
33042
14684
22026
9788
14682
5235
5584
6058
6136
6195
6524
6608
6980
7670
8260
8376

8388
8496
9204
9786
9912
10470
10620
12390

4456
5570
6684

2643
2829
2937
3524
3772
3795
3872
3916
4048
4114
4450
4600
4840
5060
5286
5658
5808
5874
6072
6900
7260
7590

2348
3522

1335
1412
1424
1472
1564
1780
1840
2118
2136
2208
2346

                    2670
                    2760
        752
        940
        1128
        1410
385
        485
        579
        595
        663
        765
                1149
                1532
                2298
        772
        776
        832
        884
        896
        952
        970
        1040
        1120
        1152
        1158
        1164
        1190
        1224
        1248
        1326
        1344
        1428
        1440
        1530
        1560
        1680
    429
465
        699
        885
                1329
                    2505
                    2672
                    3340
                    4008

```
                        5010
                  1772
                  2658
            932
            944
            1180
            1398
            1416
            1770
      482
      488
      495
      496
      525
            789
            1052
            1578
      572
      574
      610
      616
      620
      650
      700
      732
      738
      744
      770
      792
      858
      900
      924
      930
      990
      1050
==========================================
```

A computer program for building rooted trees representing prime numbers was written in 1995 by V.I. Shalack (Russian Academy of Sciences). The program consists of three sub-routines:

(1) ERATOS, generating prime numbers by the sieve of Eratosthenes;

(2) INVEULER, calculating values of inverse Euler totient function $\varphi^{-1}(m)$;

(3) P_TREES, building the rooted trees.

The most difficult sub-routine is (2) since, to calculate $\varphi^{-1}(m)$, we need to know $\varphi^{-1}(x)$ for all $x < m$, and to have prime factorization of the elements of $\varphi^{-1}(x)$.[20]

With the help of the first version of V. Shalack's program, values of $\varphi^{-1}(m)$ for $m \leq 200000$ were calculated; accordingly, the program could build rooted trees for the first 729 prime numbers.[21] In August 2000, V. Shalack extended the capacity of the program by expanding the range of prime numbers for which it can build rooted trees and by incorporating into the program's functionality the calculation of the power (the number of vertices) of the generated trees. The program generated 1207706 prime numbers, which permits to calculate the values of $\varphi^{-1}(m)$ for $m \leq 3317744$. The construction of rooted trees left off at the prime number 30689 (No. 3310) since the calculations of $\varphi^{-1}(3228368)$ gave the set of values {5104689, 6053205, 6456752, 8070940, 10209378, 12106410}. Nevertheless, with the help of Rytin's program the calculation gave the following result: $\varphi^{-1}(5104688) = \{\varnothing\}$ and $\varphi^{-1}(6053204) = \{\varnothing\}$. Thus, the building of the graph for 30689 had been completed.

The irregular distribution of the power of trees representing prime numbers seems rather striking. For the majority of prime numbers, the corresponding rooted tree contains just two elements: $\{p, 2p\}$. There are, however, such "monsters" as 21089 (No. 2371), whose tree contains 5557 elements; the tree for number 30689 (No. 3310) contains 2255 elements. We conjecture that later on this discrepancy in powers of trees will only widen.

Statistic distribution of rooted trees representing prime numbers constitutes the subject of special investigation. The question as to the largest prime number involved in the calculation of values of $\varphi_l^{*-1}(p)$ deserves special consideration, too. This last question boils down to finding a function $P(p)$ generating such largest prime number. It is evident that the graph of values of $P(p)$ grows rather sharply. It turns out that, in a sense, every prime number contains "information" pertaining to, as it were, its *maximal prime companion*.

---

[20] In July 2000, the author, on the Internet, came across a program for finding all the solutions to the equation $\varphi(n) = m$ [Rytin, 1999]. The program is very useful since it immediately calculates all the values n for an arbitrary m. Consequently, a program for building of rooted trees $\mathcal{T}_p$ can be significantly simplified.

[21] In [Karpenko, 2000] Table 2 contains the values of $\varphi^{-1}(m)$ for $m \leq 5000$. It permits to build the rooted trees for the first fifty-two prime numbers.

V.3.1. Hypothesis about finiteness of rooted trees

It is worth noting that all the rooted trees for prime numbers built thus far are *finite*. Let's come back to the inverse function $\varphi_l^{*-1}(m)$, and let m be prime number 13. In our example above, the Łukasiewicz logic $Ł_{35+1}$ is transformed into precomplete logic $Ł_{13+1}$, or $\varphi_l^*(35) = 13$. In this case, $k = 3$. We will take into account only odd values of $\varphi_l^{*-1}(m)$, i.e. $\{v_0\}$. Then $\varphi_1^{*-1}(13) = \{21\}$; $\varphi_2^{*-1}(21) = \{25, 33\}$; $\varphi_3^{*-1}(25) = \{35\}$. But, now, 34 is a value of $\varphi(69)$; therefore, $\varphi_4^{*-1}(35) \neq \varnothing$. Since 68 is not a value of $\varphi(n)$, $\varphi_5^{*-1}(69) = \varnothing$. Note that $\varphi_3^{*-1}(33) = \{51\}$, and $\varphi_4^{*-1}(51) = \varnothing$.

The following question arises: Is l (the number of iterations) finite in every case? If the number of even numbers that are not the values of $\varphi(n)$ were finite, then all classes X_p starting from some prime number p would be infinite. But as we have remarked above, citing [Rehman, 1977], there exist infinitely many even numbers that are not values of $\varphi(n)$. Thus, we have discovered a necessary condition for the finiteness of every class X_p.

HYPOTHESIS 1. $\forall p(|\ \{n: \exists k(\varphi_k^*(n) = p)\}\ | < \aleph_0,$

i.e. for each prime number p its equivalence class X_p (or its rooted tree) is finite [Karpenko, 1986]. Otherwise, there exists an increasing infinite sequence of odd numbers v_{o_i} such that only the first v_{o_i} is prime and each $v_{o_i} - 1$ is a value of $\varphi(v_{o_{i+1}} - 1)$.

Please, note that there is likely to be some connection between the cardinal degrees of completeness of Łukasiewicz logics $Ł_n$ (see the section I.5 and Table 1) and the rooted trees for prime numbers, i.e. between the functions $\gamma(n)$ and $\varphi_l^{*-1}(p)$. Thus, it is easy to notice that, for *large* rooted trees, the function $\gamma(p)$ gives large values. Since the cardinal degree of completeness of a finite-valued logic $Ł_n$ is always finite, any link found between the above-mentioned functions would give confirmation to the above Hypothesis 1.

Summing up, every prime number p can be represented in the form of a rooted tree with a distinguished vertex p (its root). Also, we have got a structuralization of prime numbers; this, however, is only the very beginning.

V.4.   p-Abelian groups

Rooted trees are ubiquitous in combinatorics, computer science, chemistry, physics, and some other disciplines. Therefore, any correspondence between rooted trees and prime numbers could potentially prove very useful. In this respect, it is worth noting the paper [Hales, 1971], in which it is showed that every rooted tree can be used to define an *Abelian p-group*.

Let T be a rooted tree. If (i, j) is an edge of T, and if j lies "between" i and t, where t is the root; we assign the direction $i \to j$ to the edge (i, j); this turns T into a directed graph.

Now, let p be a prime number. We define an Abelian group G_p by postulating its generators and equations, as follows: take one generator x_i for each vertex i (other than t) of T; and for each directed edge $i \to j$ of T, stipulate the equation $px_i = x_j$ (or, if $j = t$, the equation $px_i = 0$). Then, G_p is a p-primary Abelian group, i.e. the elements of G_p are ordered according to the powers of p. Since p in G_p is an arbitrary prime number, then giving the representation of T in the form of G_p is of purely theoretical interest. But, as in our case every prime number p is represented in the form of only "its" rooted tree $\mathcal{T}_p$, we can now construct a p-Abelian group for every such tree.

Example. Let p be 3. Then, to the rooted tree $\mathcal{T}_3$ (see above) – which has, apart from its root only two vertices: $x_1 = 4$ и $x_2 = 6$ – there corresponds the following p-Abelian group G_3:

$$x_1 \oplus x_1 \oplus x_1 = 0 \text{ и } x_2 \oplus x_2 \oplus x_2 = 0,$$

where $\oplus$ is a commutative and associative binary operator with the identity element 0; G_3 has 9 ($=3^2$) elements. In similar vein, every prime number has a corresponding G_p-structure.

[Hales, 1971] also introduced an equivalence relation (*similarity*) on rooted trees and posed a hard enumeration problem connected with this relation. The problem was solved by P. Schultz [Schulz, 1982] with the help of a program enumerating all presentations of a given group. As a result, we have the following picture:

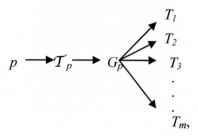

$$p \longrightarrow \mathcal{T}_p \longrightarrow G_p$$

$$T_1$$
$$T_2$$
$$T_3$$
$$\vdots$$
$$T_m,$$

Figure 2.

where p is a prime number; $\mathcal{T}_p$ is a rooted tree representing p, and G_p is a p-Abelian group representing $\mathcal{T}_p$; rooted trees $\{T_1, T_2, T_3, \ldots, T_m\}$ are all presentations of G_p, or the similarity class of the rooted tree $\mathcal{T}_p$. Then, it follows that there exist prime numbers representing some similarity class of rooted trees although for many rooted trees such a similarity class consists of a single rooted tree – for example, $\{\mathcal{T}_2\}$, $\{\mathcal{T}_{11}\}$, $\{\mathcal{T}_{23}\}$, and so on.

V.4.1.   Carmichael's totient function conjecture

It is well worth stressing that our correspondence between prime numbers and rooted trees is not a 1-1-corespondence. For example, it is impossible to encode by a prime number the following rooted tree:

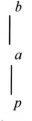

$$b$$

$$a$$

$$p$$

Figure 3.

Since vertex a must be denoted by an odd number v_0 (because of vertex b), there must exist some vertex $2a$ (because of multiplicity of the function $\varphi(n)$: $\varphi(2 \cdot v_0) = \varphi(2) \cdot \varphi(v_0) = 1 \cdot \varphi(v_0) = \varphi(v_0)$), i.e. the tree should then have the following form:

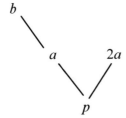

Figure 4.

This "new" tree, however, does not encode a prime number, either, since to allow only edge directed to a, *Carmichael's totient function conjecture* has to be confirmed. The conjecture is a well-known unsolved problem concerning possible values of the function $A(m)$ calculating the number of solutions of the equation $\varphi(x) = m$, also called the *multiplicity* of m.[22] Carmichael conjectured in 1907 (see [Carmichael, 1922] that, for every m, the equation $\varphi(x) = m$ has either none or at least two solutions. In other words, no totient function can have the multiplicity of 1. For example, $A(14)$ is 0, while $A(10)$ is 2: $\varphi(x) = 10$ has two solutions, 11 and 22. Carmichael's conjecture is true if and only if there exist such $m \neq n$ such $\varphi(n) = \varphi(m)$ ([Ribenboim, 1996, pp. 39-40]). Although it is still an open problem, very large lower bounds for a possible counterexample are relatively easy to obtain, the latest being $10^{10^{10}}$ ([Ford, 1988]). This means that, in that interval, the equation $\varphi(x) = a$ has, if any, at least two solutions: b_1 and b_2; then, Figure 3 should look like this:

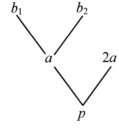

Figure 5.

---

[22] Note that, in 1950's, W. Serpinski conjectured that, for every integer $k \geq 2$, there exists a number m for which the equation $\varphi(x) = m$ has exactly k solutions. P. Erdös proved that if some multiplicity occurs once, it occurs infinitely often [Erdös, 1958]. Serpinski's conjecture was confirmed by K. Ford [Ford, 1999].

However, no prime number is the first one hundred does not have such a tree.

As our correspondence between prime numbers and rooted trees is not 1-1-corespondence[23], it follows that there can exist equivalence classes $\{T_1, T_2, T_3, \ldots, T_m\}$ not containing a tree $\mathcal{T}_p$ for any p. Then, the following problem arises:

Problem. Describe a class of p-Abelian groups characterized only by $\mathcal{T}_p$-trees.

Once the solution for this problem is found, it might lead on to confirming Carmichael's conjecture provided the class of p-Abelian groups in question enjoys the properties not allowing to represent the groups in this class in the shape depicted in Fig. 2 and Fig. 3. We do know, however, these trees are encoded by prime numbers if $\mathcal{T}_p$-trees can be represented as *canceled* $\mathcal{T}_p$-trees.

V.5. Canceled rooted trees

Note that the rooted tree for the prime number 241 has the vertices for which $\varphi_i^{*-1}(v_0) \neq \varnothing$, i.e. the vertices encoded by odd numbers v_0 such that $\varphi(x) = v_0$-1 (such vertices are printed in bold in our computer-terminal type representation of the tree). We call any tree with such vertices a *canceled rooted tree* (or, a CRT for short).

In what follows, we give CRTs for some prime numbers. For the *majority* of prime numbers, a CRT contains only one vertex, denoted by the prime number itself. For example, in the first hundred of natural numbers we have CRTs denoted by

2, 3, 5, 11, 17, 19, 23, 29, 31, 47, 53, 59, 67, 71, 79, 83, 89, 97.

CRTs for some other prime numbers look like this:

---

[23] F. Göbel [Göbel, 1980] established a 1-1-corespondence between rooted trees and natural numbers.

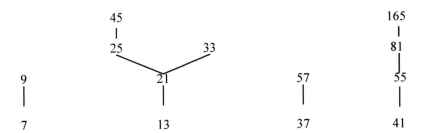

Figure 6.

Further on, for the prime numbers 101 (№ 26) to 541 (№ 100), we will represent CRTs as on a computer terminal.

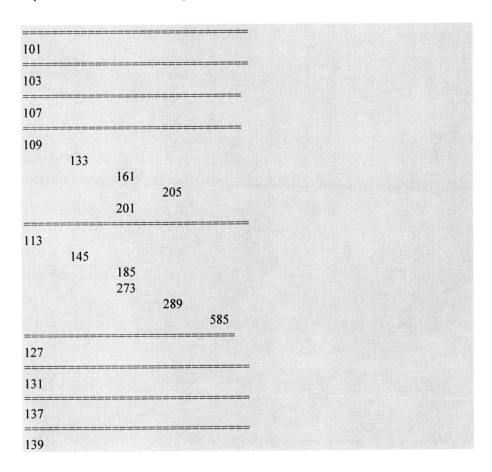

149

151

157
169
261
393

163

167

173

179

181
209
265
217
333
501
625
689
865
1377
1665
785
985
1113
1185
297

191

193
221
253
301
453
381
441
357
537

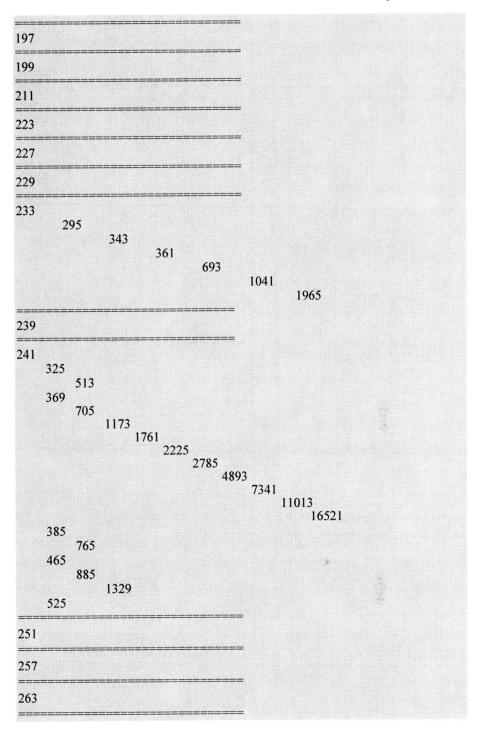

197

199

211

223

227

229

233
295
343
361
693
1041
1965

239

241
325
513
369
705
1173
1761
2225
2785
4893
7341
11013
16521

385
765
465
885
1329
525

251

257

263

===================================

269

===================================

271

===================================

277

     329
     417

===================================

281

===================================

283

===================================

293

===================================

307

===================================

311

===================================

313

    477
        717
    507
       529
          801
              1025
                  1285
              1353
          897
          1005

===================================

317

===================================

331

===================================

337

===================================

347

===================================

349

===================================

353

    445

===================================

===
359
===
367
===
373
===
379
===
383
===
389
===
397
      469
           553
      621
           933
                1401
===
401
      505
           637
===
409
===
419
===
421
      633
===
431
===
433
      481
           861
                1293
                    1941
                        2913
      545
           685
      657
      665
      777
===
439

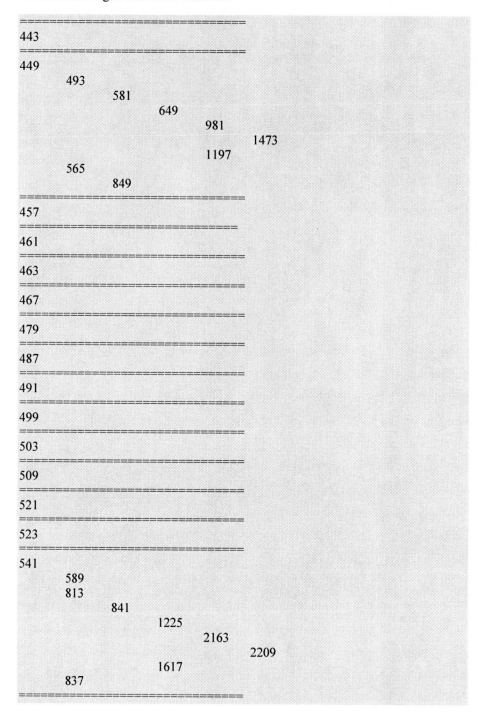

443

449
493
581
649
981
1473
1197
565
849

457

461

463

467

479

487

491

499

503

509

521

523

541
589
813
841
1225
2163
2209
1617
837

Now, it is not hard to show that the trees on the Figure 2 and Figure 3, considered as CRTs, are encoded, for example, by prime numbers 401 and 1381:

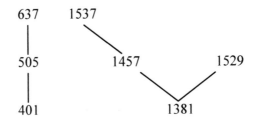

Figure 7.

The representation of prime numbers in the form of CRTs poses some problems, however. Let T_r be a number of rooted trees with r vertices. The problem as to the number of rooted trees of order r is considered in detail in [Harary, 1994], which provides rooted trees for $r \leq 4$: [24]

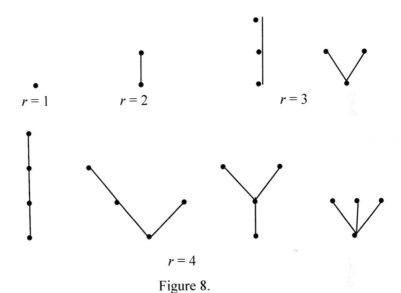

$r = 4$

Figure 8.

---

[24] The web-page [Ruskey, 1995-2003] gives rooted trees for $r = 5$. These trees are generated by a computer program.

It can be shown that for every rooted tree on the Fig. 6, considered as a CRT, there exists a corresponding prime number. All CRTs of order 5 are also encoded by prime numbers.

The numbers for rooted trees on vertices for $n = 1, 2, 3, \ldots$ are 1, 1, 2, 4, 9, 20, 48, 115, 286, 719, 1842, 4766...

HYPOTHESIS 2. *For every r ($r \in N$), each CRT is encoded by some prime number.*

In connection with the representation of prime numbers in the form of rooted trees, the following questions arise:

1. Is it possible to have $\mathcal{T}_p$-trees with an arbitrary finite number of vertices n, $n \geq 2$? For example, for number 2, there exists a $\mathcal{T}_2$-tree, along with other trees, such as a $\mathcal{T}_{11}$-tree, a $\mathcal{T}_{23}$-tree and so on; for number 3, we have a $\mathcal{T}_3$-tree; for number 4, a $\mathcal{T}_5$-tree; for number 5, a $\mathcal{T}_{457}$-tree; for number 6, a $\mathcal{T}_{17}$-trees; on the other hand, there is no $\mathcal{T}_p$-tree for number 7 among the first hundred of prime numbers.

2. A similar question arises for CRTs with $n \geq 1$. For example, for number 1, we have a $\mathcal{T}_2$-tree, along with other trees: a $\mathcal{T}_3$-tree, a $\mathcal{T}_5$-tree, and so on; for number 2, we have a $\mathcal{T}_7$-tree; for 3, a $\mathcal{T}_{277}$-tree; for 4, a $\mathcal{T}_{41}$-tree; for 5, $\mathcal{T}_{13}$-tree; for 6, a $\mathcal{T}_{43}$-tree; for 7, a $\mathcal{T}_{73}$-tree, and so on.

3. The question also arises as to the frequency of appearance of the powers of $\mathcal{T}_p$-trees and CRTs.

Thus, we moved from a partition of natural numbers into equivalence classes such that every class contains one and only one prime number on to another partition, defined only on prime numbers. The latter partition is induced by the equivalence relation with respect to either the number of vertices of rooted trees $\mathcal{T}_p$ or the number of vertices of CRTs. In the latter case, the number in question is identical to the number of such applications of the inverse function $\varphi_l^{*-1}(v_0)$ that $\varphi_l^{*-1}(v_0) \neq \varnothing$ in the process of building CRTs.

In Table 2 below, powers of $\mathcal{T}_p$-trees and CRTs are given for $p \leq 1000$.

VI. A matrix logic for prime numbers and the law of generation of classes of prime numbers

VI.1. Characterization of prime numbers by matrix logic K_{n+1}

Let's start by reminding that Finn's theorem giving the precompleteness criterion for sets of functions of Łukasiewicz $n+1$-valued logic $Ł_{n+1}$ (see section II.8) provides the characterization of prime numbers by classes of functions precomplete in P_{n+1}. This result led us to conceive of the idea of defining a many-valued matrix logic K_{n+1} such that it has tautologies if and only if n is a prime number. Then, prime numbers could be characterized by classes of tautologies of K_{n+1}. The first publication expounding this idea appeared as [Karpenko, 1982].

Let's define, in a usual way, the matrix $\mathfrak{M}_{n+1}^K$:

$$\mathfrak{M}_{n+1}^K = < V_{n+1}, \sim, \to^K, \{n\} > \quad (n \geq 3, n \in N), \text{ where}$$

$$\sim x = n\text{-}x,$$

$$x \to^K y = \begin{cases} y, \text{ if } 0 < x < y < n \text{ and } (x, y) \neq 1 & (i) \\ y, \text{ if } 0 < x = y < n & (ii) \\ x \to y \text{ otherwise} & (iii), \end{cases}$$

where $(x, y) \neq 1$ means that x and y are not relatively prime numbers, and $x \to y$ is a Łukasiewicz implication. Let's note, for the sake of comparing $x \to^K y$ and $x \to y$, that $x \to y$ can be defined as follows:

$$x \to y = \begin{cases} n, \text{ if } x \leq y \\ n - x + y, \text{ if } x > y. \end{cases}$$

Thus, $x \to^K y$ significantly differs from $x \to y$ when $0 < x \leq y < n$.

We denote by K_{n+1} the set of all functions of $\mathfrak{M}_{n+1}^K$ definable as superpositions of $\sim x$ and $x \to^K y$.

For us, it is important that the primitive connectives $\sim x$ and $x \to y$ of $Ł_{n+1}$ enjoy the following two properties:

$$\sim\sim x = x \qquad \text{(the law of double negation)},$$

$$x \to y = \sim y \to \sim x \qquad \text{(the law of contraposition)}.$$

In our further discussion, we will draw on the following properties of divisibility relation (p.d.r. for short):

(I p.d.r.). If x and y are divided by z, then $x + y$ is also divided by z.

(II p.d.r.). If x and y are divided by z and $x \geq y$, then $x-y$ is also divided by z.

Lemma 1. *Let n be a prime number. If $x < n-x$, then $x \to^K \sim x = n$.*

Proof.

Let's begin by showing that $(x, n-x) = 1$, i.e. x and $n-x$ are relatively prime numbers. Assume, for the sake of contradiction, that $(x, n-x) \neq 1$. Then, $d|x$ and $d|n-x$, where d is a divisor of x and $n-x$ other than 1. Therefore, it follows from (I p.d.r.) that $d|(x+n-x)$, i.e. $d|n$. But this contradicts the assumption of the lemma stating that n is a prime number. Thus, $(x, n-x) = 1$. Hence, in virtue of (*iii*) of the definition of $x \to^K y$, we have $x \to^K \sim x = n$.

Now, we can give a definition of prime numbers in terms of classes of tautologies of $\mathbf{K_{n+1}}$.

Theorem 1. *For any $n \geq 3$, n is a prime number iff $n \in K_{n+1}$.*

Proof.

I. *Sufficiency*: if n is a prime number, then $n \in K_{n+1}$. Let n be a prime number. We show that the following formula U, then, takes on the value n:

$$\sim((x \to^K y) \to^K \sim(x \to^K y)) \to^K (\sim(x \to^K y) \to^K (x \to^K y)),$$

Consider the subformulas $U_1 = (x \to^K y) \to^K \sim(x \to^K y)$ and $U_2 = \sim(x \to^K y) \to^K (x \to^K y)$ of U. Evidently, when $x \to^K y = 0$ or $x \to^K y = n$, $U = n$. In virtue of Lemma 1, if $x \to^K y < n/2$, then $U_1 = n$ and $\sim U_1 = 0$. Hence, in virtue of the definition of $x \to^K y$ (*iii*), $\sim U_1 \to^K U_2 = n$, and consequently, $U = n$. If $x \to^K y > n/2$, then $U_2 = n$. Hence $\sim U_1 \to^K U_2 = n$ and consequently, $U = n$.

II. *Necessity*: if $n \in K_{n+1}$, then n is a prime number. We will prove this by contraposition. If n is not prime, it has at least one divisor other than 1 and n. Let d be such a divisors of n, and let D be the set of elements $m \cdot d$ with $m = 1$, 2, ..., $(n/d)-1$. We show that D is closed with respect to $\sim x$ and $x \to^K y$.

Let $x \in D$, i.e. $x = m \cdot d$. Then $\sim x = n-(m \cdot d)$. It then follows from (II p.d.r.) that $d|n-(m \cdot d)$. Hence, $\sim x \in D$.

Let $x, y \in D$ and $x = m_i \cdot d$, $y = m_j \cdot d$. Then $x \to^K y = m_i \cdot d \to^K m_j \cdot d$. There are two cases to consider.

1. $m_i \leq m_j$. By definition, $x \to^K y = m_i \cdot d \to^K m_j \cdot d = m_j \cdot d$. Hence, $x \to^K y \in D$.

2. $m_i > m_j$. By definition, $x \to^K y = m_i \cdot d \to^K m_j \cdot d = n - m_i \cdot d + m_j \cdot d$. It then follows from (II p.d.r.) and (I p.d.r.) that $d | (n - m_i \cdot d + m_j \cdot d)$. Hence, $x \to^K y \in D$.

Therefore, there is no superposition $f(x)$ of functions $\sim x$ and $x \to^K y$ such that $f(x) = n$ if n is not a prime number.

Thus, Theorem 1 provides a new definition of prime numbers. Considering, in a usual way, a propositional language SL and a valuation function v from SL into M_{n+1}^K (see section III.2), we obtain that $\mathbf{K_{n+1}}$ has tautologies if and only if n is a prime number, i.e. every prime numbers is determined by the corresponding class of tautologies. In this connection, there arises the question of the functional properties of $\mathbf{K_{n+1}}$.

VI.2.   Functional properties of $\mathbf{K_{n+1}}$

Theorem 2. *For any $n \geq 3$ such that n is a prime number, $K_{n+1} = Ł_{n+1}$.*

Proof.

I. $K_{n+1} \subseteq Ł_{n+1}$.

This follows from the definition of $x \to^K y$ together with either Theorem 6 from section IV.10 or from McNaughton's criterion (section IV.3).

II. $Ł_{n+1} \subseteq K_{n+1}$.

To prove this inclusion, we have to define $x \to y$ through superposition of $\sim x$ and $x \to^K y$. It can be done with the help of the following sequence of definitions taken from [Karpenko, 1982]:

(A) $x \to^1 y = \sim((y \to^K x) \to^K \sim(y \to^K x)) \to^K (x \to^K y)$

(B) $x \vee^1 y = (x \to^1 y) \to^1 y$

(C) $x \to^2 y = ((x \to^K y) \to^K (\sim y \to^K \sim x)) \vee^1$
$\qquad ((\sim y \to^K \sim x) \to^K ((x \to^K y))$

(D) $x \vee^K y = (x \to^K y) \to^K y$

(E) $x \vee^2 y = (x \vee^K y) \vee^1 (y \vee^K x) = x \vee y = max(x, y)$

(F) $x \to^3 y = (x \to^K y) \vee^2 (\sim y \to^K \sim x)$

(G) $x \vee^3 y = (x \to^3 y) \to^3 y$

(H) $x \to^4 y = ((x \vee^3 y) \to^2 (x \vee^2 y)) \to^1 (x \to^3 y)$

$$(I) \quad x \to^5 y = (x \to^4 y) \vee^1 (\sim y \to^4 \sim x) = x \to y = min(n, n\text{-}x+y).$$

For a detailed proof of the Theorem, we have to consider the above formulae and describe those of their properties needed to define $x \to y$. Here we present a slightly corrected version of the paoer [Karpenko, 1989].

$$(A) \quad x \to^1 y = \sim((y \to^K x) \to^K \sim(y \to^K x)) \to^K (x \to^K y)$$

1. Let $x < y$ and $y = n$. Then $x \to^K y = n$ (iii). Hence, $x \to^1 y = n = x \to y$.

2. Let $x = y$. We have two cases.

2.1. $x < n/2$. There are two sub-cases to consider.

2.1.1. $x = 0$. Then, in virtue of the definition of $x \to^K y$ (iii), $x \to^K y = n$. Hence, $x \to^1 y = x \to y$.

2.1.2. $x \neq 0$. Then $x \to^1 y = \sim(x \to^K \sim x) \to^K x$. In virtue of Lemma 1, $x \to^K \sim x = n$. Hence, $\sim(x \to^K \sim x) = 0$ and, consequently, $x \to^1 y = 0 \to^K x = n = x \to y$.

2.2. $x > n/2$. There are two sub-cases to consider.

2.2.1. $x = n$. The, $x \to^K y = n$. Hence, $x \to^1 y = n = x \to y$.

2.2.2. $x \neq n$. Then $x \to^1 y = \sim(x \to^K \sim x) \to^K x = (n-(n- x + n-x)) \to^K x = (2x-n) \to^K x$. We will show that $((2x-n), x) = 1$. Assume, for the sake of contradiction, that $d \mid (2x-n)$ and $d \mid x$, where $d \neq 1$. Note that $(2x-n) < x$ for any $x > n/2$. Then, it follows from (II p.d.r.) that $d \mid (x-(2x-n))$, i.e. $d \mid (n-x)$. But from (I p.d.r.) it follows that that $d \mid (n-x + x)$, i.e. $d \mid n$, which contradicts the assumption that n is a prime number. Thus, our assumption is false, and in virtue of the definition of $x \to^K y$ (iii), $\sim(x \to^K \sim x) \to^K x = n$. Hence, $x \to^1 y = x \to y$.

3. $x > y$ and $x = n$. Then $y \to^K x = n$ and $x \to^K y = y$. Hence, $x \to^1 y = \sim(n \to^K 0) \to^K y = n \to^K y = y = x \to y$.

Notice that, unlike $x \to^K y$, the formula $x \to^1 y$ always takes on the value n if $x = y$, just as Łukasiewicz implication $x \to y$ does. Also note that from the properties of $x \to^1 y$ it follows that, when $n = 3$ or $n = 5$, we have, for any $0 \leq x$ and $y \leq 5$, that $x \to^1 y = x \to y$. However, for $n = 7$, if $x = 4$ and $y = 2$, then $x \to^1 y = 7$ while $x \to y = 5$. Thus, if $x > y$, then $x \to^1 y = x \to y$ does not hold in general.

$$(B) \quad x \vee^1 y = (x \to^1 y) \to^1 y.$$

Since disjunction $x \vee^1 y$ is defined analogously to a Łukasiewicz disjunction $x \vee y = (x \to y) \to y$, for the cases when $x \to^1 y = (x \to y)$ (which are discussed above), $x \vee^1 y = max(x, y)$.

A matrix logic for prime numbers and the law of generation of primes

(C) $x \to^2 y = ((x \to^K y) \to^K (\sim y \to^K \sim x)) \vee^1$
$$((\sim y \to^K \sim x) \to^K ((x \to^K y)).$$

Consider the sub-formulas $C_1 = ((x \to^K y) \to^K (\sim y \to^K \sim x)$ and $C_2 = (\sim y \to^K \sim x) \to^K ((x \to^K y)$.

1. $x > y$ and $y = n$. Then $(x \to^K y) = n$. Hence, $C_2 = n$ and thus $C_1 \vee^1 C_2 = n$. Consequently, $x \to^2 y = x \to y$.

2. $x = y$.

2.1. $x < n/2$.

2.1.1. $x = 0$. Then $x \to^K y = n$ and, consequently, $x \to^2 y = x \to y$ (see C.1.)

2.1.2. $x \neq 0$. Then $C_1 = x \to^K \sim x$. In virtue of Lemma 1, $x \to^K \sim x = n$. Hence $C_1 \vee^1 C_2 = n$ and, consequently, $x \to^2 y = x \to y$.

2.2. $x > n/2$.

2.2.1. $x = n$. Then $\sim y \to^K \sim x = n$. Hence, $C_1 = n$ and thus $C_1 \vee^1 C_2 = n$. Consequently, $x \to^2 y = x \to y$.

2.2.2. $x \neq n$. Then $C_1 = \sim x \to^K x = n$ and, consequently, $x \to^2 y = x \to y$ (see C.2.1.2.)

3. $x > y$ and $x, y \in (1, ..., n-1\}$. From the definition of $x \to^K y$ it follows that $x \to^K y = x \to y$. Since $x \to y = \sim y \to \sim x$, then $x \to^K y = \sim y \to^K \sim x$. Then in virtue of the definition of $x \to^K y$ (ii), $C_1 = (x \to y) \to^K (\sim y \to \sim x) = x \to y$ and $C_2 = (\sim y \to \sim x) \to^K ((x \to y) = x \to y$. Consequently, $x \to^2 y = C_1 \vee^1 C_2 = x \to y$.

Thus, if $x > y$, then (which does not hold for $x \to^1 y$) $x \to^2 y = x \to y$ for all $x, y \in (1, ..., n-1\}$.

(D) $x \vee^K y = (x \to^K y) \to^K y$.

1. $x < y$.

1.1. $x = 0$ or/and $y = n$. Then $x \to^K y = n$ and consequently $x \vee^K y = n \to^K y = y = max\ (x, y)$.

1.2. $(x, y) = 1$. Then $x \to^K y = n$ and, consequently, $x \vee^K y = n \to^K y = y = max\ (x, y)$.

1.3. $(x, y) \neq 1$. Then $x \to^K y = y$ and, consequently, $x \vee^K y = y \to^K y = y = max(x, y)$.

2. $x = y$.

2.1. $x \in \{0, n\}$. Then $x \to^K y = n$ and, consequently, $x \vee^K y = n \to^K y = y = max\ (x, y)$.

2.2. $x, y \in (1, ..., n-1\}$. Then $x \to^K y = y$ and, consequently, $x \vee^K y = y \to^K y = y = max(x, y)$.

3. $x > y$. We have two cases.

3.1. $x \neq n$. Then in virtue of the definition of $x \to^K y$, $x \vee^K y = (n-x+y) \to^K y = n-(n-x+y) = x = max(x, y)$.

3.2. $x = n$. Then $x \vee^K y = (n-n+y) \to^K y = y \to^K y$. We have two subcases.

89

3.2.1. $y = 0$. Then $x \vee^K y = 0 \to^K 0 = n = max(x, y)$.

3.2.2. $y \neq 0$. Then $x \vee^K y = y \to^K y \neq max(x, y)$. Thus, since it is not commutative, the disjunction $x \vee^K y$ differs from $x \vee y$. Hence $x \vee^K y \neq max(x, y)$. Note that $x \vee^1 y = max(x, y)$ in the last subcase and this specific property of $x \vee^1 y$ has been used in defininig $x \to^2 y$.

(E) $x \vee^2 y = (x \vee^K y) \vee^1 (y \vee^K x) = x \vee y = max(x, y)$

It sufficies to test case (D.3.2.2). Let $x = n$ and $y \neq 0$. Then $y \vee^K x = n$ and, consequently, $x \vee^2 y = max(x, y)$.

(F) $x \to^3 y = (x \to^K y) \vee^2 (\sim y \to^K \sim x)$.

1. $x < y$.

1.1. $x = 0$ or/and $y = n$. Then $x \to^K y = n$ and $\sim y \to \sim x = n$. Hence, $x \to^3 y = n = x \to y$.

1.2. $(x, y) = 1$ or/and $(n-y, n-x) = 1$. Then $x \to^K y = n$ or/and $\sim y \to \sim x = n$. Hence, $x \to^3 y = n = x \to y$.

1.3. $(x, y) \neq 1$ and $(n-y, n-x) \neq 1$ (for example, if $n = 11$, $x = 2$ and $y = 8$, then $\sim y = 3$ and $\sim x = 9$). Then $x \to^K y = y$ and $\sim y \to^K \sim x = \sim x$. Hence, $x \to^3 y = y \vee^2 \sim x$. We have two sub-cases.

1.3.1. $(x + y) < n$. Then $y < (n-x)$. In the opposite case, $(x + y) > n$, which contradicts the hypothesis. Consequently, $x \to^3 y = \sim x$.

1.3.2. $(x + y) > n$. Then $y > (n-x)$ and, consequently, $x \to^3 y = y$.

2. $x = y$.

2.1. $x < n/2$.

2.1.1. $x = 0$. Then $x \to^K y = n$ and, consequently, $x \to^3 y = n \vee^2 (\sim y \to^K \sim x) = n = x \to y$.

2.1.2. $x \neq 0$. Then $x \to^K y = x$ and $\sim y \to^K \sim x = \sim x$, where $x < \sim x$. Hence, $x \to^3 y = x \vee^2 \sim x = \sim x$.

2.2. $x > n/2$.

2.2.1. $x = n$. Then $\sim y \to^K \sim x = n$ and, consequently, $x \to^3 y = (x \to^K y) \vee^2 n = x \to y$.

2.2.2. $x \neq n$. Then $x \to^K y = x$ and $\sim y \to^K \sim x = \sim x$, where $x > \sim x$. Hence, $x \to^3 y = x \vee^2 \sim x = x$.

3. $x > y$. Since $x \to^K y = x \to y$ for this case, then $x \to^3 y = x \to y$.

(G) $x \vee^3 y = (x \to^3 y) \to^3 y$.

1. $x < y$.

1.2. $(x, y) = 1$ or/and $(n - y, n - x) = 1$. Then in virtue of (F.1.1), $x \to^3 y = n$ and, consequently, $x \vee^3 y = n \to^3 y = y$ (F.3).

1.3. $(x, y) \neq 1$ or/and $(n - y, n - x) \neq 1$.

1.3.1. $(x + y) < n$. Then $x \to^3 y = \sim x$ $(F.1.3.1)$ and, consequently, $x \vee^3 y = \sim x \to^3 y$. In virtue of $(F.1.3.1)$, $y < \sim x$. Hence, $x \vee^3 y = \sim x \to^3 y = n - (n - x) + y = x + y$ $(F.3)$.

1.3.2. $(x + y) > n$. Then $x \to^3 y = y$ $(F.1.3.2)$ and, consequently, $x \vee^3 y = y \to^3 y$. From conditions $(G.1)$ and $(G.1.3.2)$, it follows that $y > n/2$ and from $(G.1.3)$ it follows that $y \neq n$. Hence, $x \vee^3 y = y \to^3 y = y$ $(F.2.2.2)$.

2. $x = y$.

2.1. $x < n/2$.

2.1.1. $x = 0$. Then $x \to^3 y = n$ $(F.2.1.1)$ and, consequently, $x \vee^3 y = n \to^3 y = y$ $(F.3)$.

2.1.2. $x \neq 0$. Then $x \to^3 y = \sim x$ $(F.2.1.2)$ and, consequently, $x \vee^3 y = \sim x \to^3 y$. Since $x = y$ and $\sim x > x$, then $x \vee^3 y = \sim x \to^3 x = n - (n - x) + x = x + x = 2x$ $(F.3)$.

2.2. $x > n/2$.

2.2.1. $x = n$. Then $x \to^3 y = x$ $(F.2.2.1)$ and, consequently, $x \vee^3 y = n \to^3 y = y$ $(F.3)$.

2.2.2. $x \neq n$. Then $x \to^3 y = x$ $(F.2.2.2)$ and, consequently, $x \vee^3 y = x \to^3 y = x$ $(F.2.2.2)$.

3. $x > y$ and $x \neq n$. Then $x \to^3 y = n - x + y$ $(F.3)$. Since $(n - x + y) > y$, then $x \vee^3 y = (n - x + y) \to^3 y = n - (n - x + y) + y = x$ $(F.3)$.

(H) $x \to^4 y = ((x \vee^3 y) \to^2 (x \vee^2 y)) \to^1 (x \to^3 y)$.

1. $x < y$.

1.1. $x = 0$ or/and $y = n$. Then $x \to^3 y = n$ $(F.1.1)$. Hence, in virtue of the properties $x \to^1 y$, $x \to^4 y = n = x \to y$.

1.2. $(x, y) = 1$ or/and $(n - y, n - x) = 1$. Then $x \to^3 y = n$ $(F.1.2)$. Hence, in virtue of the properties $x \to^1 y$, $x \to^4 y = n = x \to y$.

1.3. $(x, y) \neq 1$ or/and $(n - y, n - x) \neq 1$.

1.3.1. $(x + y) < n$. Then $x \vee^3 y = x + y$ $(G.1.3.1)$, $x \vee^2 y = y$ (E) and $x \to^3 y = \sim x$ $(F.1.3.1)$. Hence, $x \to^4 y = ((x + y) \to^2 y) \to^1 \sim x = (n - x - y + y) \to^1 \sim x = \sim x \to^1 \sim x = n = x \to y$ $(C.3)$ and $(A.2.2.2)$.

1.3.2. $(x + y) > n$. Then $x \vee^3 y = y$ $(G.1.3.2)$, $x \vee^2 y = y$ (E) and $x \to^3 y = y$ $(F.1.3.2)$. Hence, $x \to^4 y = (y \to^2 y) \to^1 y = n \to^1 y = y$ $(C.2.2.2)$ and $(A.3)$. Consequently, $x \to^4 y \neq x \to y$.

2. $x = y$.

2.1. $x < n/2$.

2.1.1. $x = 0$. Then $x \to^3 y = n$ $(F.2.1.1)$. Hence, $x \to^4 y = n = x \to y$ $(H.1.1)$.

2.1.2. $x \neq 0$. Then $x \vee^3 y = 2x$ (G.2.1.2), $x \vee^2 y = x$ (E) and $x \to^3 y = \sim x$ (F.2.1.2). Hence, $x \to^4 y = (2x \to^2 x) \to^1 \sim x = (n - 2x + x) \to^1 \sim x = \sim x \to^1 \sim x = n = x \to y$ (C.3) and (A.2.2.2).

2.2. $x > n/2$.

2.2.1. $x = n$. Then $x \to^3 y = n$ (F.2.2.1). Hence, $x \to^4 y = n = x \to y$ (H.1.1).

2.2.2. $x \neq n$. Then $x \vee^3 y = x$ (G.2.2.2), $x \vee^2 y = x$ (E) and $x \to^3 y = x$ (F.2.2.2). Hence $x \to^4 y = (x \to^2 x) \to^1 x = n \to^1 x = x$ (C.2.2.2) and (A.3). Consequently, $x \to^4 y \neq x \to y$.

3. $x > y$.

3.1. $x \neq n$. Then $x \vee^3 y = x$ (G.3), $x \vee^2 y = x$ (E) and $x \to^3 y = x \to y$ (F.3). Hence, $x \to^4 y = (x \to^2 x) \to^1 (x \to y) = n \to^1 (x \to y) = x \to y$ (C.2) and (A.3).

3.2. $x = n$. Then $x \vee^2 y = n$ (E) and $x \to^3 y = y$ (F.3). Hence, $(x \vee^3 y) \to^2 (x \vee^2 y) = n$ (in virtue of the properties $x \to^2 y$). Consequently, $x \to^4 y = n \to^1 y = y = x \to y$ (A.3).

$$(I) \quad x \to^5 y = (x \to^4 y) \vee^1 (\sim y \to^4 \sim x) = x \to y = min(n, n - x + y).$$

Consider the cases, when $x \to^4 y = x \to y$ holds. Since $x \to y = \sim y \to \sim x$, then $x \to^4 y = \sim y \to^4 \sim x$. Hence, in virtue of the properties $x \vee^1 y$, $x \to^5 y = x \to^4 y = x \to y$.

Consider the two cases from (H), in which $x \to^4 y \neq x \to y$ holds.

1.3.2. $x < y$, $(x, y) \neq 1$ or/and $(n - y, n - x) \neq 1$, $(x + y) > n$. Then $x \to^4 y = y$ (H.1.3.2) and $\sim y \to^4 \sim x = n$ (H.1.3.1). Hence, $x \to^5 y = y \vee^1 n = n = x \to y$ (B).

2.2.2. $x = y$, $x > n/2$ $x \neq n$. Then $x \to^4 y = x$ (H.2.2.2) and $\sim y \to^4 \sim x = n$ (H.2.1.2). Hence, $x \to^5 y = x \vee^1 n = n = x \to y$ (B).

Thus, for any x and y, $x \to^5 y = x \to y$ and, consequently, $Ł_{n+1} \subseteq K_{n+1}$. This finishes the proof of Theorem 2.[25]

We can now give yet another definition of prime numbers, in terms of the equality of two classes of functions.

Theorem 3. For any $n \geq 3$, n is a prime number iff $K_{n+1} = Ł_{n+1}$.

I. If $n \geq 3$ is a prime number, then $K_{n+1} = Ł_{n+1}$. This is the claim of Theorem 2.

II. If $K_{n+1} = Ł_{n+1}$, then $n \geq 3$ is a prime number. We prove this by contraposition. Let $n \geq 3$ be not a prime number. Then it follows from Theorem 1 (necessity) that $n \notin K_{n+1}$. It, furthermore, follows from the

---

[25] Note that this Theorem holds good for the case $n = 2$. For example, in this case we can let $1 \to^K 1 = n$.

properties of $Ł_{n+1}$ that $n \in Ł_{n+1}$ for any $n \geq 3$. Consequently, if $n \geq 3$ is not a prime number, then $K_{n+1} \neq Ł_{n+1}$.

VI.3.   Matrix logic K'_{n+1}

The proof of Theorem 2 is rather involved. We can, however, simplify it by defining another matrix logic for prime numbers.

Let's define the matrix $\mathfrak{M}_{n+1}^{K'}$ in the following way:

$$\mathfrak{M}_{n+1}^{K'} = \,<V_{n+1}, \sim, \to^{K'}, \{n\}> \quad (n \geq 3, n \in N), \text{ where}$$

$$\sim x = n\text{-}x,$$

$$x \to^{K'} y = \begin{cases} x, \text{ if } 0 < x < y < n, (x,y) \neq 1 \text{ and } (x+y) \leq n & (i_1) \\ y, \text{ if } 0 < x < y < n, (x,y) \neq 1 \text{ and } (x+y) > n & (i_2) \\ y, \text{ if } 0 < x = y < n & (ii) \\ x \to y \text{ otherwise} & (iii), \end{cases}$$

where $(x, y) \neq 1$ means that x and y are not relatively prime numbers, and $x \to y$ is a Łukasiewicz implication.

Thus, case (i) in the definition of $x \to^K y$ is divided into two sub-cases, (i_1) and (i_2), in the definition of $x \to^{K'} y$.

We denote by K'_{n+1} the set of all function of $\mathfrak{M}_{n+1}^{K'}$ definable as superpositions of $\sim x$ and $x \to^{K'}y$.

Lemma 1'. *Let n be a prime number. If $x < n\text{-}x$, then $x \to^{K'} \sim x = n$.*

The proof is analogous to the proof of Lemma 1.

Theorem 1'. *For any $n \geq 3$, n is a prime number iff $n \in K'_{n+1}$.*

The proof is analogous to the proof of Theorem 1.

Theorem 2'. *For any $n \geq 3$ such that n is a prime number, $K'_{n+1} = Ł_{n+1}$.*

We are only interested in the following inclusion:

II. $Ł_{n+1} \subseteq K'_{n+1}$:

$$(A') \ x \to^{1'} y = \sim((y \to^{K'} x) \to^{K'} \sim(y \to^{K'} x)) \to^{K'} (x \to^{K'} y)$$

$$(B') \ x \vee^{1'} y = (x \to^{1'} y) \to^{1'} y$$

$$(C') \ x \to^{2'} y = \sim y \to^{K'} \sim x$$

$$(D') \ x \to^{S} y = x \to^{2'} ((y \to^{2'} y) \to^{2'} \sim y$$

$$(E') \ x \to^{3'} y = \sim y \to^{S} \sim x$$

$$(F') \ x \to^{4'} y = ((x \to^{\kappa'} y) \to^{3'} (\sim y \to^{\kappa'} \sim x)) \vee^{1'}$$

$$((\sim y \to^{\kappa'} \sim x) \to^{3'} (x \to^{\kappa'} y)) = x \to y.$$

The full proof using this sequence of definitions can be found in [Karpenko, 1995]. There follows a simplified proof, drawing on this simplified sequence of definitions:

$$(A') \ x \to^{1'} y = \sim((y \to^{\kappa'} x) \to^{\kappa'} \sim(y \to^{\kappa'} x)) \to^{\kappa'} (x \to^{\kappa'} y)$$

$$(B') \ x \vee^{1'} y = (x \to^{1'} y) \to^{1'} y$$

$$(C') \ x \to^{2'} y = \sim(\sim x \to^{\kappa'} \sim(x \to^{\kappa'} x)) \to^{\kappa'} y$$

$$(D') \ x \to^{3'} y = ((x \to^{\kappa'} y) \to^{2'} (\sim y \to^{\kappa'} \sim x)) \vee^{1'}$$

$$(\sim y \to^{\kappa'} \sim x) \to^{2'} (x \to^{\kappa'} y)) = x \to y.$$

The testing of the formulae (A') and (B') is analogous to the proof of Theorem 1. Let us consider the formulae (C') and (D').

$$(C') \ x \to^{2'} y = \sim(\sim x \to^{\kappa'} \sim(x \to^{\kappa'} x)) \to^{\kappa'} y.$$

We denote the subformula $\sim(\sim x \to^{\kappa'} \sim(x \to^{\kappa'} x))$ of (C') by X.

1. $y = n$. Then, $X \to^{\kappa'} y = n$ (*iii*). Hence, $x \to^{2'} y = n$.

2. $y = \sim x$ and $x < \sim x$. Then, $X = x$. Hence, $x \to^{2'} y = x \to^{\kappa'} \sim x$, and in virtue Lemma 1', $x \to^{2'} y = n = x \to y$.

3. $x = y$.

3.1. $x = 0$. Then, $X = \sim(n \to^{\kappa'} \sim(0 \to^{\kappa'} 0)) = n$. Hence, $x \to^{2'} y = n \to^{\kappa'} 0 = 0$.

3.2. $0 < x = y < n$. Then, $X = \sim(\sim x \to^{\kappa'} \sim x) = x$. Hence, $x \to^{2'} y = x \to^{\kappa'} y = x$.

3.3. $x = n$. Then, $x \to^{2'} y = X \to^{\kappa'} n = n$.

Thus, if $x = y$, the function $x \to^{2'} y$ is *idempotent* for all x. This property will prove very useful later on.

$$(D') \ x \to^{3'} y = ((x \to^{\kappa'} y) \to^{2'} (\sim y \to^{\kappa'} \sim x)) \vee^{1'}$$

$$(\sim y \to^{\kappa'} \sim x) \to^{2'} (x \to^{\kappa'} y)) = x \to y.$$

Let $D'_1 = (x \to^{\kappa'} y) \to^{2'} (\sim y \to^{\kappa'} \sim x)$ and $D'_2 = (\sim y \to^{\kappa'} \sim x) \to^{2'} (x \to^{\kappa'} y)$.

1. $x < y$.

A matrix logic for prime numbers and the law of generation of primes

1.1. $x = 0$ or/and $y = n$. Then $x \to^{K'} y = n$ (iv). Hence, $D'_2 = n$ (C'.1) and $D'_1 \vee^{1'} n = n$. Consequently, $x \to^{3'} y = n = x \to y$.

1.2. $(x, y) = 1$ and/or $(n\text{-}y, n\text{-}x) = 1$. Then $x \to^{K'} y = n$ and/or $\sim y \to^{K'} \sim x = n$, by (iv). Hence, $D'_1 = n$ and/or $D'_2 = n$ (C'.1). Then $D'_1 \vee^{1'} D'_2 = n$. Consequently, $x \to^{3'} y = n = x \to y$.

1.3. $(x, y) \neq 1$ and $(n\text{-}y, n\text{-}x) \neq 1$. There are two sub-cases to consider.

1.3.1. $(x + y) < n$. Then in virtue of the definition of $x \to^{K'} y$ (i_1), $x \to^{K'} y = x$. Evidently, if $(x + y) < n$, then $(n\text{-}y + n\text{-}x) > n$. Hence, $\sim y \to^{K'} \sim x = \sim x$ (i_2). Since $x < \sim x$, then $D'_1 = x \to^{2'} \sim x = n$ (C'.2). So $n \vee^{1'} D'_2 = n$. Consequently, $x \to^{3'} y = n = x \to y$.

1.3.2. $(x + y) > n$. Then in virtue of the definition of $x \to^{K'} y$ (i_2), $x \to^{K'} y = y$. Evidently, if $(x + y) > n$ then $(n\text{-}y + n\text{-}x) < n$. Hence, $\sim y \to^{K'} \sim x = \sim y$ (i_1). Since $\sim y < y$, we have $D'_2 = \sim y \to^{2'} y = n$ (C'.2). So, $D'_1 \vee^{1'} n = n$. Consequently, $x \to^{3'} y = n = x \to y$.

2. $x = y$.

2.1. $x < n/2$.

2.1.1. $x = 0$. Then $x \to^{K'} y = n$ (iii). Hence, $D'_2 = n$ (C'.1) and $D'_1 \vee^{1'} n = n$. Consequently, $x \to^{3'} y = n = x \to y$.

2.1.2. $x \neq 0$. Then $x \to^{K'} y = x$ and $\sim y \to^{K'} \sim x = \sim x$ (ii). Hence, $D'_1 = x \to^{2'} \sim x = n$ (C'.2) and $n \vee^{1'} D'_2 = n$. Consequently, $x \to^{3'} y = n = x \to y$.

2.2. $x > n/2$.

2.2.1. $x = n$ (see (D'.2.1.1)).

2.2.2. $x \neq n$. Then $x \to^{K'} y = x$ and $\sim y \to^{K'} \sim x = \sim x$ (ii). Hence, $D'_2 = \sim x \to^{2'} x = n$ (C'.2) and $D'_1 \vee^{1'} n = n$. Consequently, $x \to^{3'} y = n = x \to y$.

3. $x > y$. Then $x \to^{K'} y = x \to y$ and $\sim y \to^{K'} \sim x = \sim y \to \sim x$ (iii). Since $x \to y = \sim y \to \sim x$, we have $D'_1 = x \to y$ and $D'_2 = x \to y$ (C'.3). So, $D'_1 \vee^{1'} D'_2 = x \to y$. Consequently, $x \to^{3'} y = x \to y$.

Thus, for any x and y, $x \to^{3'} y = x \to y$ and, consequently, $Ł_{n+1} \subseteq K'_{n+1}$. This finishes the proof of Theorem 2'.

Theorem 3'. *For any $n \geq 3$, n is a prime number iff $K'_{n+1} = Ł_{n+1}$.*

The proof is analogous to the proof of Theorem 3.

From Theorem 3 and Theorem 3' we obtain the following important corollary:

Corollary 1. *For any $n \geq 3$, n is a prime number iff $K_{n+1} = K'_{n+1}$.*

Thus, when n is a prime number, functions $x \to^{K} y$ and $x \to^{K'} y$ are identical.

Let us note that the definition of $\rightarrow$ given via A through I includes 21 345 281 occurrences of $\rightarrow^K$; on the other hand, the definition of $\rightarrow$ given via A' through D' includes only 167 occurrences of $\rightarrow^{K'}$ and 113 occurrences of negation $\sim$. Probably, the latter is the shortest definition of $x \rightarrow y$. Then, the following question arises: can we replace these functions by a single function? In other words, does there exist a Sheffer stroke for the set of functions $\{\sim x, x \rightarrow^{K'} y\}$? (This problem was raised for the set $\{\sim x, x \rightarrow^K y\}$ in [Karpenko, 1989, p. 474]).

VI.4. Sheffer stroke for prime numbers

Let us replace in the formula

$$(C')\, x \rightarrow^{2'} y = \sim(\sim x \rightarrow^{K'} \sim(x \rightarrow^{K'} x)) \rightarrow^{K'} y,$$

introduced in the previous section as a definition of a kind of implication, every variable by its negation and, afterwards, replace every occurrence x by an occurrence of y, and vice versa. Let's denote the resultant formula by (S):

$$(S)\, x \rightarrow^{s} y = \sim(y \rightarrow^{K'} \sim(\sim y \rightarrow^{K'} \sim y)) \rightarrow^{K'} \sim x.$$

Let's consider some properties of the function $x \rightarrow^{s} y$.

1. $x = 0$. Then in virtue of definition of $\sim x$, $\sim x = n$ and, consequently, $x \rightarrow^{s} y = n$ (iii).

2. $0 < x, y < n$. Then in virtue of definition of $x \rightarrow^{K'} y$ (ii) and the law of double negation, $x \rightarrow^{s} y = \sim y \rightarrow^{K'} \sim x$.

3. $x = n$. Let 3.1. $0 < y < n$. Then $x \rightarrow^{s} y = \sim y \rightarrow^{K'} 0 = y$.

4. $y = 0$. Then $0 \rightarrow^{K'} \sim x = n$.

5. $y = n$. Then $x \rightarrow^{s} y = n \rightarrow^{K'} \sim x = \sim x$.

From (5) it follows that $n \rightarrow^{s} n = 0$. Hence, the matrix definition of $x \rightarrow^{s} y$ looks like this:

$$x \rightarrow^{s} y = \begin{cases} n, & \text{if } y = 0 \\ \sim x, & \text{if } y = n \\ \sim y \rightarrow^{K} \sim x, & \text{if } 0 < x, y < n \\ x \rightarrow y & \text{otherwise.} \end{cases}$$

It is worth comparing the above definition with the definition of a Sheffer stroke for $Ł_{n+1}$ in section IV.4. It is also worth bearing in mind that the contraposition $x \to y = \sim y \to \sim x$ holds in $Ł_{n+1}$ (see Corollary 2 below), but it does not hold in K'_{n+1} or K_{n+1}.

Let S_{n+1} denote the set of all superpositions of the function $x \to^s y$, i.e. $S_{n+1} = [x \to^s y]$.

Theorem 4. *For any $n \geq 3$ such that n is a prime number, $S_{n+1} = K'_{n+1}$* ([Karpenko, 1994]).

Proof.

I. $S_{n+1} \subseteq K'_{n+1}$.

$$x \to^s y = \sim(y \to^{K'} \sim(\sim y \to^{K'} \sim y)) \to^{K'} \sim x.$$

II. $K'_{n+1} \subseteq S_{n+1}$.

All we have to do is define the functions $\sim x$ and $x \to^{K'} y$ through $x \to^s y$.

(*a*) $\sim x = x \to^s x$.

1. $x = 0$. Then $x \to^s x = n$ (*S.1*).
2. $0 < x < n$. Then $x \to^s x = \sim x \to^{K'} \sim x$ (*S.2*). Hence, $x \to^s x = \sim x$ (*ii*).
3. $x = n$. Then $x \to^s x = n \to^s n = 0$ (*S.5*).

Thus, for any x, $\sim x = x \to^s x$.

(*b*) $n = \sim(x \to^s \sim x) \to^s \sim(\sim x \to^s x)$.

Let us denote formula (*b*) by N, and let $N_1 = \sim(x \to^s \sim x)$ and $N_2 = \sim(\sim x \to^s x)$.

1. $x < n/2$.

1.1. $x = 0$. Then $N_1 = \sim(0 \to^s n)$. Since $0 \to^s n = n$ (*S.1*), $N_1 = 0$. Hence, $N = 0 \to^s N_2 = n$ (*S.1*).

1.2. $x \neq 0$. Then in virtue of the definition of $x \to^s y$ (*S.2*) and the law of double negation, $N_1 = \sim(x \to^{K'} \sim x)$. In virtue of Lemma 1', $x \to^{K'} \sim x = n$. Hence, $N_1 = 0$, and consequently, $N = 0 \to^s N_2 = n$ (*S.1*).

2. $x > n/2$.

2.1. $x = n$. Then $N_2 = \sim(0 \to^s n) = 0$ (see 1.1 above). Hence, $N = N_1 \to^s 0 = n$ (*S.4*).

2.2. $x \neq n$. $N_2 = \sim(\sim x \to^{K'} x) = 0$ (see 1.2 above). Hence, $N_2 = 0$, and consequently, $N = N_1 \to^s 0 = n$ (*S.4*).

Thus, $N = n$, for any x.

(*c*) $x \to^{K'} y = \sim y \to^s (n \to^s \sim x)$.

1. $x = 0$. Then $x \to^{K'} y = n$ (*iii*), and $\sim y \to^{S} (n \to^{S} \sim x) = \sim y \to^{S} 0 = n$ (*S.5* and *S.4*). Hence, the equality (*c*) does hold.

2. $0 < x, y < n$. Then $\sim y \to^{S} (n \to^{S} \sim x) = \sim y \to^{S} \sim x$ (*S.3.1*). In virtue of (*S.2*), $x \to^{S} y = \sim y \to^{K'} \sim x$. Hence, $\sim y \to^{S} \sim x = \sim\sim x \to^{K'} \sim\sim y = x \to^{K'} y$.

3. $x = n$. Then $x \to^{K'} y = y$ (*iii*) and $\sim y \to^{S} (n \to^{S} \sim x) = \sim y \to^{S} n = \sim\sim y = y$ (*S.4* and *S.5*).

4. $y = 0$. Then $x \to^{K'} y = \sim x$ and $\sim y \to^{S} (n \to^{S} \sim x) = n \to^{S} \sim x = \sim x$ (*S.3.1* applied twice).

5. $y = n$. Then $x \to^{K'} y = n$ and $\sim y \to^{S} (n \to^{S} \sim x) = 0 \to^{S} \sim x = n$ (*S.3.1* and *S.1*).

Hence, (*c*) holds and, consequently, $x \to^{S} y$ is a Sheffer stroke for K'_{n+1}. In virtue of Corollary 1, the function $x \to^{S} y$ is also a Sheffer stroke for K_{n+1}.

From Theorems 4 and 2' there follows, by transitivity, the following corollary.

Corollary 2. *For any $n \geq 3$ such that n is a prime number, $S_{n+1} = Ł_{n+1}$.*

Theorem 5. *For any $n \geq 3$, n is a prime number iff $S_{n+1} = Ł_{n+1}$.*

The proof is analogous to the proof of Theorem 3 in section VI.2.

From Theorems 5 and 3' there follows, by transitivity, the following corollary.

Corollary 3. *For any $n \geq 3$, n is a prime number iff $S_{n+1} = K'_{n+1}$.*

From Corollaries 1 and 3 there follows, by transitivity, the following corollary.

Corollary 4. *For any $n \geq 3$, n is a prime number iff $S_{n+1} = K_{n+1}$.*

From Corollary 2 and McKinsey's result on Sheffer stroke for $Ł_{n+1}$, there follows the following corollary.

Corollary 5. *For any $n \geq 3$ such that n is a prime number $S_{n+1} = E_{n+1}$.*

Theorem 6. *For any $n \geq 3$, n is a prime number iff $S_{n+1} = E_{n+1}$.*

The proof is analogous to the proof of Theorem 3 in section VI.2.

Thus, we have given the characterization of prime numbers in terms of a Sheffer stroke alone. It is worth noticing that the definition of the Sheffer stroke $x \to^{S} y$ for K'_{n+1} is constant for any prime n: the definition of $x \to^{S} y$ contains exactly three occurrences implication $x \to^{K'} y$ and exactly five occurrences of negation $\sim x$. By contrast, the definition of the Sheffer stroke $x \to^{E} y$ for $Ł_{n+1}$ depends on n.

One more remark is in order here. If we replace all occurrences of $x \rightarrow^{\kappa'} y$ and $\sim x$ in $(A') - (D')$ with their Sheffer stroke definition, the number of occurrences of $x \rightarrow^{s} y$ in $(A') - (D')$ (see section VI.3) will become 648 042 744 959. The number of occurrences of $x \rightarrow^{s} y$ in $(A) - (I)$ (see section VI.2) will be astronomical. This might well be the lengthiest formula encountered the logical literature.

VI.5.   The law of generation of classes of prime numbers

The equation $K_{n+1} = K'_{n+1}$ (see Corollary 1 above) suggest the following idea: let's substitute, in the proof of Theorem 2', $x \rightarrow^{\kappa} y$ for $x \rightarrow^{\kappa'} y$ and denote the resulting sequence of formulae by $(A^*) - (D^*)$. It is not difficult to show that the formula (D^*):

$$x \rightarrow^* y = ((x \rightarrow^{\kappa} y) \rightarrow^2 (\sim y \rightarrow^{\kappa} \sim x)) \vee^1$$
$$(\sim y \rightarrow^{\kappa} \sim x) \rightarrow^2 (x \rightarrow^{\kappa} y))^{26}$$

defines Łukasiewicz implication $x \rightarrow y$ only for the first five odd prime numbers: 3, 5, 7, 11 and 13. However, if $n = 17$, $x = 2$, and $y = 12$, $x \rightarrow^* y = 15$, while $x \rightarrow y = 17$. One can show that a sequence of iterations based on (D^*) will induce classes of prime numbers, in the following way: each formula of the sequence, obtained by some iterative operation from (D^*) and denoted by $\mathcal{D}_i$ ($i = 1, 2, 3,...$) will define the implication $x \rightarrow y$ of L_{n+1} for some prime ns – this primes will form a class determined by $\mathcal{D}_i$. Now, we describe that sequence. Let

$$A_0 = ((x \rightarrow^{\kappa} y) \rightarrow^2 (\sim y \rightarrow^{\kappa} \sim x)) \text{ and}$$
$$B_0 = ((\sim y \rightarrow^{\kappa} \sim x) \rightarrow^2 ((x \rightarrow^{\kappa} y)).$$

Now,

$$\mathcal{D}_0 = A_0 \vee^1 B_0,$$
$$\mathcal{D}_1 = (A_0 \rightarrow^2 B_0) \vee^1 (B_0 \rightarrow^2 A_0),$$
$$\mathcal{D}_2 = ((A_0 \rightarrow^2 B_0) \rightarrow^2 (B_0 \rightarrow^2 A_0)) \vee^1 ((B_0 \rightarrow^2 A_0) \rightarrow^2 (A_0 \rightarrow^2 B_0))$$

and so on.

Thus, the sequence of iterations is built as follows: we substitute $\rightarrow^2$ for $\vee^1$ in the original formula $\mathcal{D}_0$ (let's denote this operation by $[\rightarrow^2/\vee^1]$) and then

---

[26] We want to draw the reader's attention to the difference between this formula and formula (C) of section VI.2. With $x = n$ and $y = 0$, $x \rightarrow^2 y = n$, while here $x \rightarrow^2 y = 0 = x \rightarrow y$ because this function is idempotent.

apply the operation of conversion (*CON*) interchanging the consequent with the antecedent of the implication resulting from the application of $[\rightarrow^2/\vee^1]$ and, lastly, links the two thus obtained formulae with $\vee^1$. In general, the iteration works thus:

$$\mathcal{D}_i = ([\rightarrow^2/\vee]\mathcal{D}_{i-1}) \vee^1 (CON([\rightarrow^2/\vee]\mathcal{D}_{i-1})).$$

To simplify the calculations, due to formula (*E*) of section IV.1.1 one can substitute the function $x \vee y = max(x, y)$ for of $x \vee^1 y$ in the definition of $\mathcal{D}_i$.

Now, let P_i denote the class of prime numbers for which $\mathcal{D}_i = x \rightarrow y$. Then

$$P_0 = \{3, 5, 7, 11, 13\},$$

$$P_1 = P_0 \cup \{17, 19\},$$

$$P_2 = P_1 \cup \{23, 29, 31, 41, 43, 53, 59, 61\}.[27]$$

With the help of a computer program developed by V.I. Shalack in 1995 we can calculate other P_i:

$$P_3 = P_2 \cup \{37, 47, 109\}.$$

Thus, P_4 contains 51 "new" prime numbers, and P_5 contains 21 more "new" prime numbers. These classes had appeared in print for the first time in [Karpenko, 1995, p. 308].

Thus, the combination of two definitions of prime numbers by logic K'_{n+1} and logic K_{n+1} gives us a very interesting result.

Now, let's take a look at the following diagrams. The rows contain a number of iterations; the columns, prime numbers.

---

[27] These three classes were first discovered in 1982.

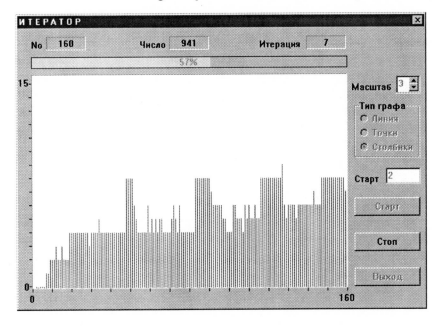

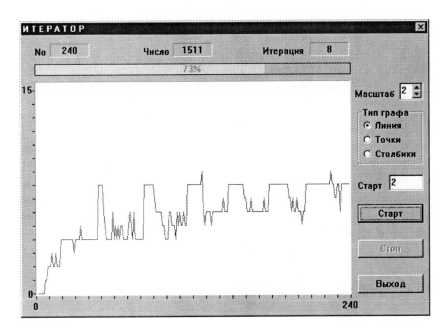

We are interested here in the partition of prime numbers into the classes P_i. Let's introduce, to this end, the function i giving, for each prime number p, the number of iterations $i(p)$ in the sequence of $\mathcal{D}_i$'s described above. The values of $i(p)$ for $p \leq 1000$ are given in Table 3.

We have already showed that $\mathcal{D}_i$ generates classes of prime numbers; the question, however, arises whether all prime numbers are thus generated? The following theorem, first formulated in [Karpenko, 1996] (see also [Karpenko, 1997a) answers the question.

Theorem 7. *Every odd prime number is contained in some class P_i.*

Proof [Karpenko, 1997b, p.178].

Let $x < y$ and $(x, y) \neq 1$, $(n-y, n-x) \neq 1$. The functions $x \to^2 y$ and $x \vee y$ are such that with the increment of i in γ_i, i.e. with the increment of the number of iterations, the values of $x \to^2 y$ and $x \vee y$ are not decremented. This is so because $x \to^2 y$, when $x < y$, takes on the value of $max(x, y)$; and when $x > y$, $x \to^2 y$ is nothing else but Łukasiewicz implication $x \to y$, i.e. $x \to^2 y = p$-x+y, which means that y gets incremented. The increment of values of x and y can not go on for ever since the number of truth values, determined by p, is finite and can not get decremented, as we have just shown; the values can, however, loop around, i.e. starting from some i, all iterations give $\mathcal{D}_i = z$, where $z \neq p$. This can happen if $\qquad (a)\ x \to^2 y = x$ and $x > y$

and, moreover,

$$(b)\ \text{if } (x, y) \neq 1, \text{ then } y \to^2 x = x.$$

It then follows that there exist such p that they are not members of any class P_i. We will show that conditions (a) and (b) are incompatible and, hence, the last statement is false.

Condition (a) holds when $x = p$-k and $y = p$-$2k$. Then, $x \to^2 y = (p$-$k \to^2 p$-$2k) = p$-$(p$-$k)$+p-$2k = p$-k. We show that $(p$-k, p-$2k) = 1$, i.e. p-k, and p-$2k$ are relatively prime numbers. For the sake of contradiction, assume otherwise, i.e. $d | p$-k и $d | p$-$2k$, where $d \neq 1$. Then, it follows from (II p.d.r.) (see section VI.1) that $d | ((p$-$k)$-$(p$-$2k))$, i.e. $d | k$. Since $d | p$-k, it follows from (I p.d.r.) that $d | (p$-k+$k)$, i.e. $d | p$, which is impossible since p is prime. Thus, $(p$-k, p-$2k) = 1$. Then, $y \to^2 x \neq x$ and, consequently, if (a) holds, (b) must fail. Therefore, for any even prime number p, after a finite number i of iterations, a class P_i will be found.

Thus, Theorem 7 is proved.

HYPOTHESIS 3. Every class P_i is finite.

In fact, formulas $\mathcal{D}_i$ can be viewed as *the law of prime numbers generation*, or more precisely, *the law of classes of prime numbers generation*. Clearly, in virtue of Corollary 1 and Theorem 4, the law can be described with Sheffer stroke $x \rightarrow^s y$ alone.

Formulae for prime numbers and functions that generate prime numbers are discussed in [Wilf, 1982] and [Ribenboim, 1997], respectively. As an example of this line of work, let's note that, in 1977, Y.V. Matiyasevicz (for details, see [Matiyasevicz, 1993]) discovered that prime numbers can be enumerated by a polynomial with 10 variables.

Let's note some irregularity in the distribution of prime numbers among classes P_i. Thus, the prime number 223 is a member of the class P_8, while the class P_5 already contains the prime number 757 (which is the greatest member of P_5). On the other hand, the powers of P_i (starting from $i = 0$) looks as follows: 5, 2, 8, 3, 51, 21, 54, 19, ... , and P_8 has more than 400 members.

The noted irregularity indicates the significant complexity of distribution of prime numbers in a natural series. In this respect, L. Euler is reported to have remarked: '...we have reason to believe that it is a mystery into which the human mind will never penetrate' (see [Ayoub, 1963, p. 37])

VII. Characterization of classes of natural numbers by logical matrices

VII.1. Prime numbers

In the preceding section, classes of prime numbers were characterized with the help of Łukasiewicz $n+1$-valued matrix ($n \in N, n \geq 2$):

$$\mathfrak{M}_{n+1}^{L} = <V_{n+1}, \sim, \to, \{1\} >,$$

where $V_{n+1} = \{0, 1, 2, ..., n\}$, $\sim$ (negation) is a unary, and $\to$ (implication) a binary, function, defined on V_{n+1} as follows:

$$\sim x = n - x,$$

$$x \to y = \begin{cases} n, \text{ if } x \leq y \\ n - x + y, \text{ if } x > y; \end{cases}$$

$\{1\}$ is the set of designated elements of $\mathfrak{M}_{n+1}^{L}$.

The set of functions of $\mathfrak{M}_{n+1}^{L}$ generated by superposition of $\sim x$ and $x \to y$ is denoted by $Ł_{n+1}$.

For the sake of comparison with subsequent characterizations of classes of natural numbers, we recall here the definition of the logical matrix $\mathfrak{M}_{n+1}^{K'}$ as well as the main results concerning that matrix:

$$\mathfrak{M}_{n+1}^{K'} = <V_{n+1}, \sim, \to^{K'}, \{n\} > \quad (n \geq 3, n \in N),$$

$$\sim x = n - x,$$

$$x \to^{K'} y = \begin{cases} x, \text{ if } 0 < x < y < n, (x, y) \neq 1 \text{ and } (x + y) \leq n & (i_1) \\ y, \text{ if } 0 < x < y < n, (x, y) \neq 1 \text{ and } (x + y) > n & (i_2) \\ y, \text{ if } 0 < x = y < n & (ii) \\ x \to y \text{ otherwise} & (iii), \end{cases}$$

where $(x, y) \neq 1$ means that x and y are not relatively prime numbers, and $x \to y$ is a Łukasiewicz implication.

The set of all functions of $M_{n+1}^{K'}$ generated by superposition of $\sim x$ and $x \to^{K'} y$ is denoted by K'_{n+1}.

Lemma 1'. *Let n be a prime number. If $x < n-x$, then $x \to^{K'} \sim x = n$.*

Theorem 1'. For any $n \geq 3$, n is a prime number iff $n \in K'_{n+1}$.

Theorem 2'. *For any $n \geq 3$ such that n is a prime number, $K'_{n+1} = Ł_{n+1}$.*

Theorem 3'. *For any $n \geq 3$, n is a prime number iff $K'_{n+1} = Ł_{n+1}$.*

Now we move on to a similar characterization of others interesting subsets of natural numbers.

VII.2. Powers of primes

In December 1981, during a discussion following the author's talk at The All-Union Institute of Science-Technical Information in Moscow on the characterization of prime numbers through Łukasiewicz $n+1$-valued logical matrices, V.K. Finn raised the question of a similar characterization of powers of primes, i.e. such numbers n that $n = p^\beta$, where p is a prime number and β a positive integer.

A clue to answering this question is provided by Finn's theorem concerning the representation of functions in $Ł_{n+1}$ ([Bochvar and Finn, 1972, Theorem 4]): each function $f \in Ł_{n+1}$ not equal to the constant 0 is definable by a superposition of $x \vee y = max(x, y)$, $x \wedge y = min(x, y)$, J-functions, and I-functions, if and only if $n = p^\beta$ (the definitions of J-functions and I-functions were given in section II). Such superposition is an analogue of the full disjunctive normal form for two-valued logic. Starting off from Finn's theorem, we can generalize the logic $\mathbf{K'}_{n+1}$ to the case of $n = p^\beta$.

To this end, the condition $(x, y) \neq 1$ in (i_1) and (i_2) of the definition of $x \to^{K'} y$ must be restricted: *there is no such divisor d $(d \neq 1)$ among common divisors of x and y that it itself or some power thereof is the only divisor of n.* Let's denote this restriction by $(i_1^{F}$ and $i_2^{F})$; otherwise, $x \to^{K'} y = n$. In turn, this restriction can be extended to the case of $0 < x = y < n$ and $x + y = n$, permitting to characterize numbers of the form 2^n. Let's denote the latter restriction by (ii^{F}).

We denote the so defined function by $x \to^{F} y$. Thus, we have the following matrix:

$$\mathfrak{M}_{n+1}^{F} = \, < V_{n+1}, \sim, \to^{F}, \{n\} > .$$

The set of functions of $\mathfrak{M}^F_{n+1}$ generated by superposition of $\sim x$ and $x \to^F y$ is denoted by F_{n+1}.

Lemma 1^F. *Let n be a power of a prime number, i.e. $n = p^\beta$. If $x < \sim x$, then $x \to^F \sim x = n$.*

Let d, a natural number different from 1 and n, be the least common divisor of x and n-x. As d is prime, let $d = p^*$, where p^* is a prime number. Suppose, for the sake of contradiction, that p^* is different from p of the statement of the lemma. Then, in the virtue of (I p.d.r.), it follows that $d^*|(x+n-x)$, i.e. $d^*|n$. But this contradicts the assumption that n is p^β. Hence $p^* = p$, and consequently, in virtue of (i_1^F and i_2^F), $x \to^F \sim x = n$.

Theorem 1. For any $n \geq 3$, $n = p^\beta$ iff $p^\beta \in F_{n+1}$.

The proof is analogous to the proof of Theorem 1 of section VI.1.

Theorem 2. *For any $n \geq 3$ such that $n = p^\beta$, $F_{n+1} = L_{n+1}$.*

I. $F_{n+1} \subseteq L_{n+1}$.

This follows from McNaughton's criterion of definability of functions in Łukasiewicz matrices $\mathfrak{M}^L_{n+1}$ (see section IV.3).

II. $L_{n+1} \subseteq F_{n+1}$.

(A^F) $x \to^1 y = \sim((y \to^F x) \to^F \sim(y \to^F x)) \to^F (x \to^F y)$

(B^F) $x \vee^1 y = (x \to^1 y) \to^1 y$

(C^F) $x \to^2 y = \sim((\sim x \to^F \sim x) \to^F \sim(x \to^F x)) \to^F y$

(D^F) $x \to^3 y = ((x \to^F y) \to^2 (\sim y \to^F \sim x)) \vee^1$
$((\sim y \to^F \sim x) \to^2 (x \to^F y)) = x \to y$.

In (A^F), we are interested in the following case:

2. $x = y$.

2.1. $(x, n) = 1$. Here, the proof is analogous to the proof of item $(A.2)$ of Theorem 2 from section VI.2, with the use of Lemma 1 being replaced by the use of Lemma 1^F.

2.2. $(x, n) \neq 1$.

2.2.1. $0 < x < n/2$. Then $x \to^F \sim x = n$ (i_1^F and i_2^F). In virtue of the definition of $x \to^F y$ (iii), $x \to^F y = n$. Hence, $\sim(x \to^F \sim x) = 0$, and consequently, $x \to^1 y = n$.

2.2.2. $n/2 < x < n$. Then $x \to^1 y = (2x-n) \to^F x$, where $(2x-n) < n$ (see $(A.2.2)$ of Theorem 1). Then, in virtue of the definition of $x \to^F y$ (i_1^F u i_2^F), $\sim(x \to^F \sim x) \to^F x = n$. Consequently, $x \to^1 y = n$.

2.3. $(x, n) \neq 1$ and $x + x = n$. Then $x \to^F y = n$ (ii^F), and consequently, $x \to^1 y = n$.

Thus, $x \to^F y$ always takes on the value n if $x = y$, just as the Łukasiewicz implication $x \to y$.

Note that the formula (C^F) differs from the formula (C') of section IV.2. It permit to verify the case $n = 2^n$. Let us check the formula

$$(C^F) \quad x \to^2 y = x \to^2 y = \sim((\sim x \to^F \sim x) \to^F \sim(x \to^F x)) \to^F y.$$

We only have to consider item 3, given the restriction (ii^F). Let us denote the subformula $\sim((\sim x \to^F \sim x) \to^F \sim(x \to^F x))$ by X^F.

3. $x = y$.

3.1. $x = 0$. Then $X^F = \sim((n \to^F n) \to^F \sim(0 \to^F 0)) = n$. Hence, $x \to^2 y = n \to^F 0 = 0$.

3.2. $0 < x = y < n$.

3.2.1. $x + y \neq n$. Then $X^F = \sim(\sim x \to^F \sim x) = x$. Hence, $x \to^2 y = x \to^F y = x$.

3.2.2. $x + y = n$.

3.2.2.1. If the restriction (ii^F) does not hold, this case is analogous to 3.2.1.

3.2.2.2. If (ii^F) holds, $X^F = \sim(n \to^F (\sim n)) = n$. Hence, $x \to^2 y = n \to^F y = y$.

Note that if we still had to work with (C') rather than (C^F), we would then have $X^F = \sim(\sim x \to^F 0) = \sim x$; then, $x \to^{2'} y = \sim x \to^F x$, i.e. the function $x \to^{2'} y$ would not be idempotent.

3.3. $x = n$. Then, $x \to^2 y = X^F \to^F n = n$.

From Theorems 1 and 2 as well as the properties of $Ł_{n+1}$ (see the proof of Theorem 3 of section VI.2), there follows the following theorem.

Theorem 3. *For any $n \geq 3$, n is p^β iff $F_{n+1} = Ł_{n+1}$.*

VII.3.   Even numbers

The problem of characterization of even numbers by logical matrices was first investigated in [Karpenko, 1999]. Here we present a slightly corrected version of that work. Let's consider the following matrix:

$$\mathfrak{M}^{e}_{n+1} = <V_{n+1}, \sim, \rightarrow^{e}, \{n\}> \quad (n \geq 3, n \in N), \text{ where}$$

$$\sim x = n - x,$$

$$x \rightarrow^{e} y = \begin{cases} x, \text{ if } 0 < x < y < n \ u \ x + y < n & (i) \\ x, \text{ if } 0 < x < y < n, \ x + y = n \text{ and } x, y \\ \quad \text{ are different } \bmod ulo \ 2 & (ii) \\ y, \text{ if } 0 < x < y < n \text{ and } x + y > n & (iii) \\ y, \text{ if } 0 < x = y < n \text{ and } x + y \neq n & (iv) \\ x \rightarrow y \text{ otherwise} & (v). \end{cases}$$

The set of functions of $\mathfrak{M}^{e}_{n+1}$ generated by superposition of $\sim x$ and $x \rightarrow^{e} y$ is denoted by E_{n+1}.

Lemma 1^{e}. *Let n be an even number. If $x < \sim x$, then $x \rightarrow^{e} \sim x = n$.*

Since n is an even number, x and $\sim x$ are equivalent modulo 2; their sum is n. Hence, (*ii*) in the definition of $x \rightarrow^{e} y$ does not apply, and consequently $x \rightarrow^{e} y = n$ (*v*).

Theorem 4. *For any $n \geq 2$, n is even iff $n \in E_{n+1}$.*

I. *Sufficiency*: if n even, then $n \in E_{n+1}$. Let n be even. Then, the formula U (see Theorem 1 in section VI.1):

$$\sim((x \rightarrow^{e} y) \rightarrow^{e} \sim(x \rightarrow^{e} y)) \rightarrow^{e} (\sim(x \rightarrow^{e} y) \rightarrow^{e} (x \rightarrow^{e} y)),$$

holds, i.e. $U = n$. Consider the subformulas $U_1 = (x \rightarrow^{e} y) \rightarrow^{e} \sim(x \rightarrow^{e} y)$ and $U_2 = \sim(x \rightarrow^{e} y) \rightarrow^{e} (x \rightarrow^{e} y)$ of U.

Clearly, when $x \rightarrow^{K} y = 0$ and $x \rightarrow^{K} y = n$, $U = n$. Let $x \rightarrow^{e} y < n/2$. Then, in virtue of Lemma 1^{e}, $U_1 = n$, and $\sim U_1 = 0$. Hence, in virtue of the definition of $\sim x$ and $x \rightarrow^{e} y$ (*v*), $\sim U_1 \rightarrow^{e} U_2 = 0 \rightarrow^{e} U_2 = n$, and consequently, $U = n$. Let $x \rightarrow^{e} y = n/2$. Then, in virtue of the definition of $\sim x$ and $x \rightarrow^{e} y$ (*v*), $U = n/2 \rightarrow^{e} n/2 = n$. Let $x \rightarrow^{e} y > n/2$. Then, $U_2 = n$. Hence, $\sim U_1 \rightarrow^{e} n = n$, and consequently, $U = n$.

II. *Necessity*: if $n \in E_{n+1}$, then n is even. We will prove this by contraposition. Let n be odd and $0 < x, y < n$. In virtue of the definition of $x \rightarrow^e y$, for $x \rightarrow^e y$ to be equal to n, x and y should be equivalent modulo 2. But then n has to be even. Therefore, if n is odd, $n \notin E_{n+1}$.

Theorem 5. *For any $n \geq 2$ such that n is even, $E_{n+1} = Ł_{n+1}$.*

I. $E_{n+1} \subseteq Ł_{n+1}$.

This follows from McNaughton's criterion (section IV.3).

II. $Ł_{n+1} \subseteq E_{n+1}$.

$(A^e)\ x \rightarrow^1 y = \sim((y \rightarrow^e x) \rightarrow^e \sim(y \rightarrow^e x)) \rightarrow^e (x \rightarrow^e y),$

$(B^e)\ x \rightarrow^2 y = \sim((y \rightarrow^e x) \rightarrow^e \sim(y \rightarrow^e x)) \rightarrow^e (\sim y \rightarrow^1 \sim x),$

$(C^e)\ x \vee^1 y = (x \rightarrow^2 y) \rightarrow^2 y,$

$(D^e)\ x \rightarrow^3 y = \sim((\sim x \rightarrow^e \sim x) \rightarrow^e \sim(x \rightarrow^e x)) \rightarrow^e y$

$(E^e)\ x \rightarrow^4 y = ((x \rightarrow^e y) \rightarrow^3 (\sim y \rightarrow^e \sim x)) \vee^1$
$\qquad ((\sim y \rightarrow^e \sim x) \rightarrow^3 (x \rightarrow^e y)) = x \rightarrow y.$

Let's consider the formula (A^e). We are interested in the following case:

2. $0 < x = y < n$ and $x > n/2$. Then, $(2x - n) \rightarrow^e x \neq n$. As a counterexample, take $n = 10$ and $x = 6$. Then, $x \rightarrow^1 y = 2 \rightarrow^e 6 = 2$. Therefore, to make $x \rightarrow^1 y$ equal to n for all $x = y$, the formula (B^e) is introduced. Then, $\sim y \rightarrow^1 \sim x = n$, and consequently, $x \rightarrow^2 y = n$.

From Theorems 4 and 5 as well as the properties of $Ł_{n+1}$ there follows the following theorem.

Theorem 6. *For any $n \geq 2$, n is an even number iff $E_{n+1} = Ł_{n+1}$.*

VII.4.   Odd numbers

Lastly, we provide the logical characterization of odd numbers. This is easy to do since Łukasiewicz matrices contain fixed points for negation $\sim x$. Let

$$\mathfrak{M}^o_{n+1} = <V_{n+1}, \sim, \to^o, \{n\}>, \text{ where}$$

$$\sim x = n{-}x,$$

$$x \to^o y = \begin{cases} n, \text{ if } \quad x < y & (i) \\ y, \text{ if } \ 0 < x = y < n \text{ and } x + y = n & (ii) \\ x \to y \text{ otherwise} & (iii). \end{cases}$$

The set of functions of $\mathfrak{M}^o_{n+1}$ generated by superposition of $\sim x$ and $x \to^o y$ is denoted by O_{n+1}.

Theorem 7. *For any $n \geq 2$, n is odd iff $n \in O_{n+1}$.*

I. *Sufficiency*: if n is odd then $n \in O_{n+1}$. Let n be odd. Then the following formula U^o:

$$(x \to^o y) \to^o (x \to^o y),$$

that takes on the value n.

1. Let $(x \to^o y) = 0$ and/or $(x \to^o y) = n$. Then, $U^o = n$.

2. Let $(x \to^o y) = z$ and $0 < z < n$. Since n is odd, $z + z \neq n$. and hence $z \to^o z = n$. Consequently, $U^o = n$.

II. *Necessity*: If $n \in O_{n+1}$, then n is odd. We prove this by contraposition. Let n be even. We denote the set of truth-values of the form $x + x = n$ by D^o. We will show that the set D^o is closed under $\sim x$ and $x \to^o y$.

Let $x \in D^o$. Then, $\sim x = x$, i.e. matrix $\mathfrak{M}^o_{n+1}$ contains fixed points for $\sim x$. Consequently, $\sim x \in D^o$. Let $x = y$ and $x + y = n$. Then, in virtue of the definition of $x \to^o y$, it follows that $x \to^o x = x$, i.e. in this case, the function $x \to^o y$ is idempotent. Consequently, $x \to^o x \in D^o$.

Theorem 8. *For any $n \geq 2$, such that n is odd, $O_{n+1} = Ł_{n+1}$.*

I. $O_{n+1} \subseteq Ł_{n+1}$.

This follows from McNaughton's criterion.

II. $Ł_{n+1} \subseteq O_{n+1}$.
$$x \to y = x \to^o y.$$

From Theorems 7 and 8 as well as the properties of $Ł_{n+1}$ there follows the following theorem.

Theorem 9. *For any $n \geq 2$, n is odd iff $O_{n+1} = Ł_{n+1}$.*

VII.5.   Some remarks on Goldbach conjecture

The provided characterization of even numbers through Lukasiewicz matrices is particularly thought-provoking. It seems unclear why this characterization proved to be the most difficult out of all considered in this book. It is worth noting that for proving theorems on even numbers we use, in an indirect way, a very simple arithmetical fact concerning the decomposition of an even number as a sum of two even or of two odd numbers. This makes us think of another well-known decomposition of even numbers – the one appearing in the well-known Golbbach conjecture stating that all positive even integers greater than 4 are sums of two prime numbers. The conjecture is discussed in a book by Y. Wang ([Wang, 1984]). If the conjecture is true, then for every number n, there are primes p and q such that

$$\varphi(p) + \varphi(q) = 2n,$$

where $\varphi(x)$ is the Euller's totient function ([Guy, 1994, p.105]); the totient function was described in section V.1.

E. Landau stated at the International Mathematics Congress held in Cambridge in 1912 that the Goldbach conjecture was beyond the state of the art of mathematics at the time. It is still considered to be there. A characteristic remark comes form G. Hardy, saying that "It is comparatively easy to make clever guesses; indeed there are theorems, like 'Goldbach's theorem', which have never been proved and which any fool could have guessed" [Hardy, 1999, p. 19].

Using an highly-efficient algorithm, the conjecture has been verified for all numbers up to 4×10^{14} ([Richstein, 2000]). Several workstations were required to run the program implementing the algorithm.

Nevertheless, there is no perceivable headway towards the solution of the Goldbach problem. However, it seems possible that the solution might be arrived at with the help of methods of this chapter; there follows an outline of such a possible algebra-logical approach to the problem. The items (*ii*) and (*iv*) in the definition of $x \to^e y$ must be restricted: *x and y are not prime numbers*. Otherwise $x \to^e y = n$. Let's denote the function so defined by $x \to^G y$. Then, when attempting prove an analogue of Theorem 5 (*sufficiency*) we come up against some serious difficulties, although the necessity condition is very easily proven. In fact, to define a function similar to $x \to^G y$, we need to strike a trade-off, allowing on the one hand, to represent some constant formula U and, on the other, not to invalidate the *necessity* condition. A similar trick proved to be doable in the case of even numbers.

Let's conclude by remarking, in response to the problems raised in Chapter III.5, that Łukasiewicz n-valued logics have a purely arithmetical interpretation; then, the material in this book points to some general facts concerning a large class of propositional logical systems, or even, probably, the intrigacies of the human mind (see *Concluding remarks*).

Łukasiewicz Logics and Prime Numbers

Numerical Tables

Łukasiewicz Logics and Prime Numbers

Table 1
Degrees of cardinal completeness $\gamma(Ł_n)$
(*see section* III.4)

n =	1					n =	54	=>	$\gamma(n)$ =		3
n =	2	=>	$\gamma(n)$ =		2	n =	55	=>	$\gamma(n)$ =		15
n =	3	=>	$\gamma(n)$ =		3	n =	56	=>	$\gamma(n)$ =		6
n =	4	=>	$\gamma(n)$ =		3	n =	57	=>	$\gamma(n)$ =		15
n =	5	=>	$\gamma(n)$ =		4	n =	58	=>	$\gamma(n)$ =		6
n =	6	=>	$\gamma(n)$ =		3	n =	59	=>	$\gamma(n)$ =		6
n =	7	=>	$\gamma(n)$ =		6	n =	60	=>	$\gamma(n)$ =		3
n =	8	=>	$\gamma(n)$ =		3	n =	61	=>	$\gamma(n)$ =		50
n =	9	=>	$\gamma(n)$ =		5	n =	62	=>	$\gamma(n)$ =		3
n =	10	=>	$\gamma(n)$ =		4	n =	63	=>	$\gamma(n)$ =		6
n =	11	=>	$\gamma(n)$ =		6	n =	64	=>	$\gamma(n)$ =		10
n =	12	=>	$\gamma(n)$ =		3	n =	65	=>	$\gamma(n)$ =		8
n =	13	=>	$\gamma(n)$ =		10	n =	66	=>	$\gamma(n)$ =		6
n =	14	=>	$\gamma(n)$ =		3	n =	67	=>	$\gamma(n)$ =		20
n =	15	=>	$\gamma(n)$ =		6	n =	68	=>	$\gamma(n)$ =		3
n =	16	=>	$\gamma(n)$ =		6	n =	69	=>	$\gamma(n)$ =		10
n =	17	=>	$\gamma(n)$ =		6	n =	70	=>	$\gamma(n)$ =		6
n =	18	=>	$\gamma(n)$ =		3	n =	71	=>	$\gamma(n)$ =		20
n =	19	=>	$\gamma(n)$ =		10	n =	72	=>	$\gamma(n)$ =		3
n =	20	=>	$\gamma(n)$ =		3	n =	73	=>	$\gamma(n)$ =		35
n =	21	=>	$\gamma(n)$ =		10	n =	74	=>	$\gamma(n)$ =		3
n =	22	=>	$\gamma(n)$ =		6	n =	75	=>	$\gamma(n)$ =		6
n =	23	=>	$\gamma(n)$ =		6	n =	76	=>	$\gamma(n)$ =		10
n =	24	=>	$\gamma(n)$ =		3	n =	77	=>	$\gamma(n)$ =		10
n =	25	=>	$\gamma(n)$ =		15	n =	78	=>	$\gamma(n)$ =		6
n =	26	=>	$\gamma(n)$ =		4	n =	79	=>	$\gamma(n)$ =		20
n =	27	=>	$\gamma(n)$ =		6	n =	80	=>	$\gamma(n)$ =		3
n =	28	=>	$\gamma(n)$ =		5	n =	81	=>	$\gamma(n)$ =		21
n =	29	=>	$\gamma(n)$ =		10	n =	82	=>	$\gamma(n)$ =		6
n =	30	=>	$\gamma(n)$ =		3	n =	83	=>	$\gamma(n)$ =		6
n =	31	=>	$\gamma(n)$ =		20	n =	84	=>	$\gamma(n)$ =		3
n =	32	=>	$\gamma(n)$ =		3	n =	85	=>	$\gamma(n)$ =		50
n =	33	=>	$\gamma(n)$ =		7	n =	86	=>	$\gamma(n)$ =		6
n =	34	=>	$\gamma(n)$ =		6	n =	87	=>	$\gamma(n)$ =		6
n =	35	=>	$\gamma(n)$ =		6	n =	88	=>	$\gamma(n)$ =		6
n =	36	=>	$\gamma(n)$ =		6	n =	89	=>	$\gamma(n)$ =		15
n =	37	=>	$\gamma(n)$ =		20	n =	90	=>	$\gamma(n)$ =		3
n =	38	=>	$\gamma(n)$ =		3	n =	91	=>	$\gamma(n)$ =		50
n =	39	=>	$\gamma(n)$ =		6	n =	92	=>	$\gamma(n)$ =		6
n =	40	=>	$\gamma(n)$ =		6	n =	93	=>	$\gamma(n)$ =		10
n =	41	=>	$\gamma(n)$ =		15	n =	94	=>	$\gamma(n)$ =		6
n =	42	=>	$\gamma(n)$ =		3	n =	95	=>	$\gamma(n)$ =		6
n =	43	=>	$\gamma(n)$ =		20	n =	96	=>	$\gamma(n)$ =		6
n =	44	=>	$\gamma(n)$ =		3	n =	97	=>	$\gamma(n)$ =		28
n =	45	=>	$\gamma(n)$ =		10	n =	98	=>	$\gamma(n)$ =		3
n =	46	=>	$\gamma(n)$ =		10	n =	99	=>	$\gamma(n)$ =		10
n =	47	=>	$\gamma(n)$ =		6	n =	100	=>	$\gamma(n)$ =		10
n =	48	=>	$\gamma(n)$ =		3	n =	101	=>	$\gamma(n)$ =		20
n =	49	=>	$\gamma(n)$ =		21	n =	102	=>	$\gamma(n)$ =		3
n =	50	=>	$\gamma(n)$ =		4	n =	103	=>	$\gamma(n)$ =		20
n =	51	=>	$\gamma(n)$ =		10	n =	104	=>	$\gamma(n)$ =		3
n =	52	=>	$\gamma(n)$ =		6	n =	105	=>	$\gamma(n)$ =		15
n =	53	=>	$\gamma(n)$ =		10	n =	106	=>	$\gamma(n)$ =		20

n =	107	=>	$\gamma(n)$ =	6	n =	163	=>	$\gamma(n)$ =	21
n =	108	=>	$\gamma(n)$ =	3	n =	164	=>	$\gamma(n)$ =	3
n =	109	=>	$\gamma(n)$ =	35	n =	165	=>	$\gamma(n)$ =	10
n =	110	=>	$\gamma(n)$ =	3	n =	166	=>	$\gamma(n)$ =	20
n =	111	=>	$\gamma(n)$ =	20	n =	167	=>	$\gamma(n)$ =	6
n =	112	=>	$\gamma(n)$ =	6	n =	168	=>	$\gamma(n)$ =	3
n =	113	=>	$\gamma(n)$ =	21	n =	169	=>	$\gamma(n)$ =	105
n =	114	=>	$\gamma(n)$ =	3	n =	170	=>	$\gamma(n)$ =	4
n =	115	=>	$\gamma(n)$ =	20	n =	171	=>	$\gamma(n)$ =	20
n =	116	=>	$\gamma(n)$ =	6	n =	172	=>	$\gamma(n)$ =	10
n =	117	=>	$\gamma(n)$ =	10	n =	173	=>	$\gamma(n)$ =	10
n =	118	=>	$\gamma(n)$ =	10	n =	174	=>	$\gamma(n)$ =	3
n =	119	=>	$\gamma(n)$ =	6	n =	175	=>	$\gamma(n)$ =	20
n =	120	=>	$\gamma(n)$ =	6	n =	176	=>	$\gamma(n)$ =	10
n =	121	=>	$\gamma(n)$ =	105	n =	177	=>	$\gamma(n)$ =	21
n =	122	=>	$\gamma(n)$ =	4	n =	178	=>	$\gamma(n)$ =	6
n =	123	=>	$\gamma(n)$ =	6	n =	179	=>	$\gamma(n)$ =	6
n =	124	=>	$\gamma(n)$ =	6	n =	180	=>	$\gamma(n)$ =	3
n =	125	=>	$\gamma(n)$ =	10	n =	181	=>	$\gamma(n)$ =	175
n =	126	=>	$\gamma(n)$ =	5	n =	182	=>	$\gamma(n)$ =	3
n =	127	=>	$\gamma(n)$ =	50	n =	183	=>	$\gamma(n)$ =	20
n =	128	=>	$\gamma(n)$ =	3	n =	184	=>	$\gamma(n)$ =	6
n =	129	=>	$\gamma(n)$ =	9	n =	185	=>	$\gamma(n)$ =	15
n =	130	=>	$\gamma(n)$ =	6	n =	186	=>	$\gamma(n)$ =	6
n =	131	=>	$\gamma(n)$ =	20	n =	187	=>	$\gamma(n)$ =	20
n =	132	=>	$\gamma(n)$ =	3	n =	188	=>	$\gamma(n)$ =	6
n =	133	=>	$\gamma(n)$ =	50	n =	189	=>	$\gamma(n)$ =	10
n =	134	=>	$\gamma(n)$ =	6	n =	190	=>	$\gamma(n)$ =	15
n =	135	=>	$\gamma(n)$ =	6	n =	191	=>	$\gamma(n)$ =	20
n =	136	=>	$\gamma(n)$ =	15	n =	192	=>	$\gamma(n)$ =	3
n =	137	=>	$\gamma(n)$ =	15	n =	193	=>	$\gamma(n)$ =	36
n =	138	=>	$\gamma(n)$ =	3	n =	194	=>	$\gamma(n)$ =	3
n =	139	=>	$\gamma(n)$ =	20	n =	195	=>	$\gamma(n)$ =	6
n =	140	=>	$\gamma(n)$ =	3	n =	196	=>	$\gamma(n)$ =	20
n =	141	=>	$\gamma(n)$ =	50	n =	197	=>	$\gamma(n)$ =	20
n =	142	=>	$\gamma(n)$ =	6	n =	198	=>	$\gamma(n)$ =	3
n =	143	=>	$\gamma(n)$ =	6	n =	199	=>	$\gamma(n)$ =	50
n =	144	=>	$\gamma(n)$ =	6	n =	200	=>	$\gamma(n)$ =	3
n =	145	=>	$\gamma(n)$ =	56	n =	201	=>	$\gamma(n)$ =	35
n =	146	=>	$\gamma(n)$ =	6	n =	202	=>	$\gamma(n)$ =	6
n =	147	=>	$\gamma(n)$ =	6	n =	203	=>	$\gamma(n)$ =	6
n =	148	=>	$\gamma(n)$ =	10	n =	204	=>	$\gamma(n)$ =	6
n =	149	=>	$\gamma(n)$ =	10	n =	205	=>	$\gamma(n)$ =	50
n =	150	=>	$\gamma(n)$ =	3	n =	206	=>	$\gamma(n)$ =	6
n =	151	=>	$\gamma(n)$ =	50	n =	207	=>	$\gamma(n)$ =	6
n =	152	=>	$\gamma(n)$ =	3	n =	208	=>	$\gamma(n)$ =	10
n =	153	=>	$\gamma(n)$ =	15	n =	209	=>	$\gamma(n)$ =	21
n =	154	=>	$\gamma(n)$ =	10	n =	210	=>	$\gamma(n)$ =	6
n =	155	=>	$\gamma(n)$ =	20	n =	211	=>	$\gamma(n)$ =	168
n =	156	=>	$\gamma(n)$ =	6	n =	212	=>	$\gamma(n)$ =	3
n =	157	=>	$\gamma(n)$ =	50	n =	213	=>	$\gamma(n)$ =	10
n =	158	=>	$\gamma(n)$ =	3	n =	214	=>	$\gamma(n)$ =	6
n =	159	=>	$\gamma(n)$ =	6	n =	215	=>	$\gamma(n)$ =	6
n =	160	=>	$\gamma(n)$ =	6	n =	216	=>	$\gamma(n)$ =	6
n =	161	=>	$\gamma(n)$ =	28	n =	217	=>	$\gamma(n)$ =	70
n =	162	=>	$\gamma(n)$ =	6	n =	218	=>	$\gamma(n)$ =	6

n =	219	=>	γ (n) =	6	n = 275	=> γ (n) =	6

$n = 219 \Rightarrow \gamma(n) = 6$ $n = 275 \Rightarrow \gamma(n) = 6$

$n = 220 \Rightarrow \gamma(n) = 6$ $n = 276 \Rightarrow \gamma(n) = 10$

$n = 221 \Rightarrow \gamma(n) = 50$ $n = 277 \Rightarrow \gamma(n) = 50$

$n = 222 \Rightarrow \gamma(n) = 6$ $n = 278 \Rightarrow \gamma(n) = 3$

$n = 223 \Rightarrow \gamma(n) = 20$ $n = 279 \Rightarrow \gamma(n) = 6$

$n = 224 \Rightarrow \gamma(n) = 3$ $n = 280 \Rightarrow \gamma(n) = 10$

$n = 225 \Rightarrow \gamma(n) = 28$ $n = 281 \Rightarrow \gamma(n) = 105$

$n = 226 \Rightarrow \gamma(n) = 20$ $n = 282 \Rightarrow \gamma(n) = 3$

$n = 227 \Rightarrow \gamma(n) = 6$ $n = 283 \Rightarrow \gamma(n) = 20$

$n = 228 \Rightarrow \gamma(n) = 3$ $n = 284 \Rightarrow \gamma(n) = 3$

$n = 229 \Rightarrow \gamma(n) = 50$ $n = 285 \Rightarrow \gamma(n) = 10$

$n = 230 \Rightarrow \gamma(n) = 3$ $n = 286 \Rightarrow \gamma(n) = 20$

$n = 231 \Rightarrow \gamma(n) = 20$ $n = 287 \Rightarrow \gamma(n) = 20$

$n = 232 \Rightarrow \gamma(n) = 20$ $n = 288 \Rightarrow \gamma(n) = 6$

$n = 233 \Rightarrow \gamma(n) = 15$ $n = 289 \Rightarrow \gamma(n) = 84$

$n = 234 \Rightarrow \gamma(n) = 3$ $n = 290 \Rightarrow \gamma(n) = 4$

$n = 235 \Rightarrow \gamma(n) = 50$ $n = 291 \Rightarrow \gamma(n) = 20$

$n = 236 \Rightarrow \gamma(n) = 6$ $n = 292 \Rightarrow \gamma(n) = 6$

$n = 237 \Rightarrow \gamma(n) = 10$ $n = 293 \Rightarrow \gamma(n) = 10$

$n = 238 \Rightarrow \gamma(n) = 6$ $n = 294 \Rightarrow \gamma(n) = 3$

$n = 239 \Rightarrow \gamma(n) = 20$ $n = 295 \Rightarrow \gamma(n) = 50$

$n = 240 \Rightarrow \gamma(n) = 3$ $n = 296 \Rightarrow \gamma(n) = 6$

$n = 241 \Rightarrow \gamma(n) = 196$ $n = 297 \Rightarrow \gamma(n) = 15$

$n = 242 \Rightarrow \gamma(n) = 3$ $n = 298 \Rightarrow \gamma(n) = 15$

$n = 243 \Rightarrow \gamma(n) = 10$ $n = 299 \Rightarrow \gamma(n) = 6$

$n = 244 \Rightarrow \gamma(n) = 7$ $n = 300 \Rightarrow \gamma(n) = 6$

$n = 245 \Rightarrow \gamma(n) = 10$ $n = 301 \Rightarrow \gamma(n) = 175$

$n = 246 \Rightarrow \gamma(n) = 10$ $n = 302 \Rightarrow \gamma(n) = 6$

$n = 247 \Rightarrow \gamma(n) = 20$ $n = 303 \Rightarrow \gamma(n) = 6$

$n = 248 \Rightarrow \gamma(n) = 6$ $n = 304 \Rightarrow \gamma(n) = 6$

$n = 249 \Rightarrow \gamma(n) = 15$ $n = 305 \Rightarrow \gamma(n) = 21$

$n = 250 \Rightarrow \gamma(n) = 6$ $n = 306 \Rightarrow \gamma(n) = 6$

$n = 251 \Rightarrow \gamma(n) = 15$ $n = 307 \Rightarrow \gamma(n) = 50$

$n = 252 \Rightarrow \gamma(n) = 3$ $n = 308 \Rightarrow \gamma(n) = 3$

$n = 253 \Rightarrow \gamma(n) = 175$ $n = 309 \Rightarrow \gamma(n) = 50$

$n = 254 \Rightarrow \gamma(n) = 6$ $n = 310 \Rightarrow \gamma(n) = 6$

$n = 255 \Rightarrow \gamma(n) = 6$ $n = 311 \Rightarrow \gamma(n) = 20$

$n = 256 \Rightarrow \gamma(n) = 20$ $n = 312 \Rightarrow \gamma(n) = 3$

$n = 257 \Rightarrow \gamma(n) = 10$ $n = 313 \Rightarrow \gamma(n) = 105$

$n = 258 \Rightarrow \gamma(n) = 3$ $n = 314 \Rightarrow \gamma(n) = 3$

$n = 259 \Rightarrow \gamma(n) = 20$ $n = 315 \Rightarrow \gamma(n) = 6$

$n = 260 \Rightarrow \gamma(n) = 6$ $n = 316 \Rightarrow \gamma(n) = 50$

$n = 261 \Rightarrow \gamma(n) = 50$ $n = 317 \Rightarrow \gamma(n) = 10$

$n = 262 \Rightarrow \gamma(n) = 10$ $n = 318 \Rightarrow \gamma(n) = 3$

$n = 263 \Rightarrow \gamma(n) = 6$ $n = 319 \Rightarrow \gamma(n) = 20$

$n = 264 \Rightarrow \gamma(n) = 3$ $n = 320 \Rightarrow \gamma(n) = 6$

$n = 265 \Rightarrow \gamma(n) = 105$ $n = 321 \Rightarrow \gamma(n) = 36$

$n = 266 \Rightarrow \gamma(n) = 6$ $n = 322 \Rightarrow \gamma(n) = 6$

$n = 267 \Rightarrow \gamma(n) = 20$ $n = 323 \Rightarrow \gamma(n) = 20$

$n = 268 \Rightarrow \gamma(n) = 6$ $n = 324 \Rightarrow \gamma(n) = 6$

$n = 269 \Rightarrow \gamma(n) = 10$ $n = 325 \Rightarrow \gamma(n) = 56$

$n = 270 \Rightarrow \gamma(n) = 3$ $n = 326 \Rightarrow \gamma(n) = 10$

$n = 271 \Rightarrow \gamma(n) = 105$ $n = 327 \Rightarrow \gamma(n) = 6$

$n = 272 \Rightarrow \gamma(n) = 3$ $n = 328 \Rightarrow \gamma(n) = 6$

$n = 273 \Rightarrow \gamma(n) = 21$ $n = 329 \Rightarrow \gamma(n) = 15$

$n = 274 \Rightarrow \gamma(n) = 20$ $n = 330 \Rightarrow \gamma(n) = 6$

n = 331	=>	γ (n) =	168	n = 387	=>	γ (n) =	6
n = 332	=>	γ (n) =	3	n = 388	=>	γ (n) =	10
n = 333	=>	γ (n) =	10	n = 389	=>	γ (n) =	10
n = 334	=>	γ (n) =	10	n = 390	=>	γ (n) =	3
n = 335	=>	γ (n) =	6	n = 391	=>	γ (n) =	168
n = 336	=>	γ (n) =	6	n = 392	=>	γ (n) =	6
n = 337	=>	γ (n) =	196	n = 393	=>	γ (n) =	35
n = 338	=>	γ (n) =	3	n = 394	=>	γ (n) =	6
n = 339	=>	γ (n) =	10	n = 395	=>	γ (n) =	6
n = 340	=>	γ (n) =	6	n = 396	=>	γ (n) =	6
n = 341	=>	γ (n) =	50	n = 397	=>	γ (n) =	175
n = 342	=>	γ (n) =	6	n = 398	=>	γ (n) =	3
n = 343	=>	γ (n) =	50	n = 399	=>	γ (n) =	6
n = 344	=>	γ (n) =	5	n = 400	=>	γ (n) =	20
n = 345	=>	γ (n) =	15	n = 401	=>	γ (n) =	56
n = 346	=>	γ (n) =	20	n = 402	=>	γ (n) =	3
n = 347	=>	γ (n) =	6	n = 403	=>	γ (n) =	20
n = 348	=>	γ (n) =	3	n = 404	=>	γ (n) =	6
n = 349	=>	γ (n) =	50	n = 405	=>	γ (n) =	10
n = 350	=>	γ (n) =	3	n = 406	=>	γ (n) =	21
n = 351	=>	γ (n) =	50	n = 407	=>	γ (n) =	20
n = 352	=>	γ (n) =	15	n = 408	=>	γ (n) =	6
n = 353	=>	γ (n) =	28	n = 409	=>	γ (n) =	105
n = 354	=>	γ (n) =	3	n = 410	=>	γ (n) =	3
n = 355	=>	γ (n) =	20	n = 411	=>	γ (n) =	20
n = 356	=>	γ (n) =	6	n = 412	=>	γ (n) =	6
n = 357	=>	γ (n) =	10	n = 413	=>	γ (n) =	10
n = 358	=>	γ (n) =	20	n = 414	=>	γ (n) =	6
n = 359	=>	γ (n) =	6	n = 415	=>	γ (n) =	50
n = 360	=>	γ (n) =	3	n = 416	=>	γ (n) =	6
n = 361	=>	γ (n) =	490	n = 417	=>	γ (n) =	28
n = 362	=>	γ (n) =	4	n = 418	=>	γ (n) =	6
n = 363	=>	γ (n) =	6	n = 419	=>	γ (n) =	20
n = 364	=>	γ (n) =	10	n = 420	=>	γ (n) =	3
n = 365	=>	γ (n) =	50	n = 421	=>	γ (n) =	887
n = 366	=>	γ (n) =	6	n = 422	=>	γ (n) =	3
n = 367	=>	γ (n) =	20	n = 423	=>	γ (n) =	6
n = 368	=>	γ (n) =	3	n = 424	=>	γ (n) =	10
n = 369	=>	γ (n) =	21	n = 425	=>	γ (n) =	15
n = 370	=>	γ (n) =	10	n = 426	=>	γ (n) =	10
n = 371	=>	γ (n) =	20	n = 427	=>	γ (n) =	20
n = 372	=>	γ (n) =	6	n = 428	=>	γ (n) =	6
n = 373	=>	γ (n) =	50	n = 429	=>	γ (n) =	10
n = 374	=>	γ (n) =	3	n = 430	=>	γ (n) =	20
n = 375	=>	γ (n) =	20	n = 431	=>	γ (n) =	20
n = 376	=>	γ (n) =	15	n = 432	=>	γ (n) =	3
n = 377	=>	γ (n) =	15	n = 433	=>	γ (n) =	126
n = 378	=>	γ (n) =	6	n = 434	=>	γ (n) =	3
n = 379	=>	γ (n) =	105	n = 435	=>	γ (n) =	20
n = 380	=>	γ (n) =	3	n = 436	=>	γ (n) =	20
n = 381	=>	γ (n) =	50	n = 437	=>	γ (n) =	10
n = 382	=>	γ (n) =	6	n = 438	=>	γ (n) =	6
n = 383	=>	γ (n) =	6	n = 439	=>	γ (n) =	20
n = 384	=>	γ (n) =	3	n = 440	=>	γ (n) =	3
n = 385	=>	γ (n) =	45	n = 441	=>	γ (n) =	105
n = 386	=>	γ (n) =	20	n = 442	=>	γ (n) =	20

n = 443	=>	γ (n) =	20	n = 499	=>	γ (n) =	20
n = 444	=>	γ (n) =	3	n = 500	=>	γ (n) =	3
n = 445	=>	γ (n) =	50	n = 501	=>	γ (n) =	35
n = 446	=>	γ (n) =	6	n = 502	=>	γ (n) =	6
n = 447	=>	γ (n) =	6	n = 503	=>	γ (n) =	6
n = 448	=>	γ (n) =	6	n = 504	=>	γ (n) =	3
n = 449	=>	γ (n) =	36	n = 505	=>	γ (n) =	490
n = 450	=>	γ (n) =	3	n = 506	=>	γ (n) =	6
n = 451	=>	γ (n) =	175	n = 507	=>	γ (n) =	20
n = 452	=>	γ (n) =	6	n = 508	=>	γ (n) =	10
n = 453	=>	γ (n) =	10	n = 509	=>	γ (n) =	10
n = 454	=>	γ (n) =	6	n = 510	=>	γ (n) =	3
n = 455	=>	γ (n) =	6	n = 511	=>	γ (n) =	168
n = 456	=>	γ (n) =	20	n = 512	=>	γ (n) =	6
n = 457	=>	γ (n) =	105	n = 513	=>	γ (n) =	11
n = 458	=>	γ (n) =	3	n = 514	=>	γ (n) =	15
n = 459	=>	γ (n) =	6	n = 515	=>	γ (n) =	6
n = 460	=>	γ (n) =	15	n = 516	=>	γ (n) =	6
n = 461	=>	γ (n) =	50	n = 517	=>	γ (n) =	50
n = 462	=>	γ (n) =	3	n = 518	=>	γ (n) =	6
n = 463	=>	γ (n) =	168	n = 519	=>	γ (n) =	20
n = 464	=>	γ (n) =	3	n = 520	=>	γ (n) =	6
n = 465	=>	γ (n) =	21	n = 521	=>	γ (n) =	105
n = 466	=>	γ (n) =	20	n = 522	=>	γ (n) =	3
n = 467	=>	γ (n) =	6	n = 523	=>	γ (n) =	50
n = 468	=>	γ (n) =	3	n = 524	=>	γ (n) =	3
n = 469	=>	γ (n) =	175	n = 525	=>	γ (n) =	10
n = 470	=>	γ (n) =	6	n = 526	=>	γ (n) =	50
n = 471	=>	γ (n) =	20	n = 527	=>	γ (n) =	6
n = 472	=>	γ (n) =	6	n = 528	=>	γ (n) =	6
n = 473	=>	γ (n) =	15	n = 529	=>	γ (n) =	196
n = 474	=>	γ (n) =	6	n = 530	=>	γ (n) =	4
n = 475	=>	γ (n) =	20	n = 531	=>	γ (n) =	20
n = 476	=>	γ (n) =	10	n = 532	=>	γ (n) =	10
n = 477	=>	γ (n) =	50	n = 533	=>	γ (n) =	50
n = 478	=>	γ (n) =	10	n = 534	=>	γ (n) =	6
n = 479	=>	γ (n) =	6	n = 535	=>	γ (n) =	20
n = 480	=>	γ (n) =	3	n = 536	=>	γ (n) =	6
n = 481	=>	γ (n) =	336	n = 537	=>	γ (n) =	15
n = 482	=>	γ (n) =	6	n = 538	=>	γ (n) =	6
n = 483	=>	γ (n) =	6	n = 539	=>	γ (n) =	6
n = 484	=>	γ (n) =	20	n = 540	=>	γ (n) =	10
n = 485	=>	γ (n) =	20	n = 541	=>	γ (n) =	490
n = 486	=>	γ (n) =	6	n = 542	=>	γ (n) =	3
n = 487	=>	γ (n) =	28	n = 543	=>	γ (n) =	6
n = 488	=>	γ (n) =	3	n = 544	=>	γ (n) =	6
n = 489	=>	γ (n) =	15	n = 545	=>	γ (n) =	28
n = 490	=>	γ (n) =	6	n = 546	=>	γ (n) =	6
n = 491	=>	γ (n) =	50	n = 547	=>	γ (n) =	168
n = 492	=>	γ (n) =	3	n = 548	=>	γ (n) =	3
n = 493	=>	γ (n) =	50	n = 549	=>	γ (n) =	10
n = 494	=>	γ (n) =	6	n = 550	=>	γ (n) =	10
n = 495	=>	γ (n) =	20	n = 551	=>	γ (n) =	50
n = 496	=>	γ (n) =	50	n = 552	=>	γ (n) =	6
n = 497	=>	γ (n) =	21	n = 553	=>	γ (n) =	105
n = 498	=>	γ (n) =	6	n = 554	=>	γ (n) =	6

Table 1: Degrees of cardinal completeness $\gamma(\text{Ł}_n)$

n = 555	=>	$\gamma(n) =$	6	n = 611	=>	$\gamma(n) =$ 20
n = 556	=>	$\gamma(n) =$	20	n = 612	=>	$\gamma(n) =$ 6
n = 557	=>	$\gamma(n) =$	10	n = 613	=>	$\gamma(n) =$ 175
n = 558	=>	$\gamma(n) =$	3	n = 614	=>	$\gamma(n) =$ 3
n = 559	=>	$\gamma(n) =$	50	n = 615	=>	$\gamma(n) =$ 6
n = 560	=>	$\gamma(n) =$	6	n = 616	=>	$\gamma(n) =$ 20
n = 561	=>	$\gamma(n) =$	196	n = 617	=>	$\gamma(n) =$ 105
n = 562	=>	$\gamma(n) =$	20	n = 618	=>	$\gamma(n) =$ 3
n = 563	=>	$\gamma(n) =$	6	n = 619	=>	$\gamma(n) =$ 20
n = 564	=>	$\gamma(n) =$	3	n = 620	=>	$\gamma(n) =$ 3
n = 565	=>	$\gamma(n) =$	50	n = 621	=>	$\gamma(n) =$ 50
n = 566	=>	$\gamma(n) =$	6	n = 622	=>	$\gamma(n) =$ 15
n = 567	=>	$\gamma(n) =$	6	n = 623	=>	$\gamma(n) =$ 6
n = 568	=>	$\gamma(n) =$	21	n = 624	=>	$\gamma(n) =$ 6
n = 569	=>	$\gamma(n) =$	15	n = 625	=>	$\gamma(n) =$ 196
n = 570	=>	$\gamma(n) =$	3	n = 626	=>	$\gamma(n) =$ 6
n = 571	=>	$\gamma(n) =$	168	n = 627	=>	$\gamma(n) =$ 6
n = 572	=>	$\gamma(n) =$	3	n = 628	=>	$\gamma(n) =$ 20
n = 573	=>	$\gamma(n) =$	50	n = 629	=>	$\gamma(n) =$ 10
n = 574	=>	$\gamma(n) =$	6	n = 630	=>	$\gamma(n) =$ 6
n = 575	=>	$\gamma(n) =$	20	n = 631	=>	$\gamma(n) =$ 887
n = 576	=>	$\gamma(n) =$	10	n = 632	=>	$\gamma(n) =$ 3
n = 577	=>	$\gamma(n) =$	120	n = 633	=>	$\gamma(n) =$ 15
n = 578	=>	$\gamma(n) =$	3	n = 634	=>	$\gamma(n) =$ 6
n = 579	=>	$\gamma(n) =$	10	n = 635	=>	$\gamma(n) =$ 6
n = 580	=>	$\gamma(n) =$	6	n = 636	=>	$\gamma(n) =$ 6
n = 581	=>	$\gamma(n) =$	50	n = 637	=>	$\gamma(n) =$ 50
n = 582	=>	$\gamma(n) =$	6	n = 638	=>	$\gamma(n) =$ 10
n = 583	=>	$\gamma(n) =$	20	n = 639	=>	$\gamma(n) =$ 20
n = 584	=>	$\gamma(n) =$	6	n = 640	=>	$\gamma(n) =$ 10
n = 585	=>	$\gamma(n) =$	15	n = 641	=>	$\gamma(n) =$ 45
n = 586	=>	$\gamma(n) =$	50	n = 642	=>	$\gamma(n) =$ 3
n = 587	=>	$\gamma(n) =$	6	n = 643	=>	$\gamma(n) =$ 20
n = 588	=>	$\gamma(n) =$	3	n = 644	=>	$\gamma(n) =$ 3
n = 589	=>	$\gamma(n) =$	175	n = 645	=>	$\gamma(n) =$ 50
n = 590	=>	$\gamma(n) =$	6	n = 646	=>	$\gamma(n) =$ 20
n = 591	=>	$\gamma(n) =$	20	n = 647	=>	$\gamma(n) =$ 20
n = 592	=>	$\gamma(n) =$	6	n = 648	=>	$\gamma(n) =$ 3
n = 593	=>	$\gamma(n) =$	21	n = 649	=>	$\gamma(n) =$ 126
n = 594	=>	$\gamma(n) =$	3	n = 650	=>	$\gamma(n) =$ 6
n = 595	=>	$\gamma(n) =$	105	n = 651	=>	$\gamma(n) =$ 50
n = 596	=>	$\gamma(n) =$	20	n = 652	=>	$\gamma(n) =$ 20
n = 597	=>	$\gamma(n) =$	10	n = 653	=>	$\gamma(n) =$ 10
n = 598	=>	$\gamma(n) =$	6	n = 654	=>	$\gamma(n) =$ 3
n = 599	=>	$\gamma(n) =$	20	n = 655	=>	$\gamma(n) =$ 20
n = 600	=>	$\gamma(n) =$	3	n = 656	=>	$\gamma(n) =$ 6
n = 601	=>	$\gamma(n) =$	490	n = 657	=>	$\gamma(n) =$ 21
n = 602	=>	$\gamma(n) =$	3	n = 658	=>	$\gamma(n) =$ 10
n = 603	=>	$\gamma(n) =$	20	n = 659	=>	$\gamma(n) =$ 20
n = 604	=>	$\gamma(n) =$	10	n = 660	=>	$\gamma(n) =$ 3
n = 605	=>	$\gamma(n) =$	10	n = 661	=>	$\gamma(n) =$ 887
n = 606	=>	$\gamma(n) =$	10	n = 662	=>	$\gamma(n) =$ 3
n = 607	=>	$\gamma(n) =$	20	n = 663	=>	$\gamma(n) =$ 6
n = 608	=>	$\gamma(n) =$	3	n = 664	=>	$\gamma(n) =$ 20
n = 609	=>	$\gamma(n) =$	28	n = 665	=>	$\gamma(n) =$ 15
n = 610	=>	$\gamma(n) =$	20	n = 666	=>	$\gamma(n) =$ 20

Łukasiewicz Logics and Prime Numbers

n =	667	=>	γ (n) =	50	n =	723	=>	γ (n) =	10
n =	668	=>	γ (n) =	6	n =	724	=>	γ (n) =	6
n =	669	=>	γ (n) =	10	n =	725	=>	γ (n) =	10
n =	670	=>	γ (n) =	6	n =	726	=>	γ (n) =	10
n =	671	=>	γ (n) =	20	n =	727	=>	γ (n) =	50
n =	672	=>	γ (n) =	6	n =	728	=>	γ (n) =	3
n =	673	=>	γ (n) =	336	n =	729	=>	γ (n) =	105
n =	674	=>	γ (n) =	3	n =	730	=>	γ (n) =	8
n =	675	=>	γ (n) =	6	n =	731	=>	γ (n) =	20
n =	676	=>	γ (n) =	35	n =	732	=>	γ (n) =	6
n =	677	=>	γ (n) =	20	n =	733	=>	γ (n) =	50
n =	678	=>	γ (n) =	3	n =	734	=>	γ (n) =	3
n =	679	=>	γ (n) =	20	n =	735	=>	γ (n) =	6
n =	680	=>	γ (n) =	6	n =	736	=>	γ (n) =	50
n =	681	=>	γ (n) =	105	n =	737	=>	γ (n) =	28
n =	682	=>	γ (n) =	6	n =	738	=>	γ (n) =	6
n =	683	=>	γ (n) =	20	n =	739	=>	γ (n) =	50
n =	684	=>	γ (n) =	3	n =	740	=>	γ (n) =	3
n =	685	=>	γ (n) =	175	n =	741	=>	γ (n) =	50
n =	686	=>	γ (n) =	6	n =	742	=>	γ (n) =	20
n =	687	=>	γ (n) =	15	n =	743	=>	γ (n) =	20
n =	688	=>	γ (n) =	6	n =	744	=>	γ (n) =	3
n =	689	=>	γ (n) =	21	n =	745	=>	γ (n) =	105
n =	690	=>	γ (n) =	6	n =	746	=>	γ (n) =	6
n =	691	=>	γ (n) =	168	n =	747	=>	γ (n) =	6
n =	692	=>	γ (n) =	3	n =	748	=>	γ (n) =	10
n =	693	=>	γ (n) =	10	n =	749	=>	γ (n) =	50
n =	694	=>	γ (n) =	50	n =	750	=>	γ (n) =	6
n =	695	=>	γ (n) =	6	n =	751	=>	γ (n) =	105
n =	696	=>	γ (n) =	6	n =	752	=>	γ (n) =	3
n =	697	=>	γ (n) =	105	n =	753	=>	γ (n) =	21
n =	698	=>	γ (n) =	6	n =	754	=>	γ (n) =	6
n =	699	=>	γ (n) =	6	n =	755	=>	γ (n) =	20
n =	700	=>	γ (n) =	6	n =	756	=>	γ (n) =	6
n =	701	=>	γ (n) =	175	n =	757	=>	γ (n) =	490
n =	702	=>	γ (n) =	3	n =	758	=>	γ (n) =	3
n =	703	=>	γ (n) =	105	n =	759	=>	γ (n) =	6
n =	704	=>	γ (n) =	6	n =	760	=>	γ (n) =	20
n =	705	=>	γ (n) =	36	n =	761	=>	γ (n) =	105
n =	706	=>	γ (n) =	20	n =	762	=>	γ (n) =	3
n =	707	=>	γ (n) =	6	n =	763	=>	γ (n) =	20
n =	708	=>	γ (n) =	6	n =	764	=>	γ (n) =	6
n =	709	=>	γ (n) =	50	n =	765	=>	γ (n) =	10
n =	710	=>	γ (n) =	3	n =	766	=>	γ (n) =	50
n =	711	=>	γ (n) =	20	n =	767	=>	γ (n) =	6
n =	712	=>	γ (n) =	10	n =	768	=>	γ (n) =	6
n =	713	=>	γ (n) =	15	n =	769	=>	γ (n) =	55
n =	714	=>	γ (n) =	6	n =	770	=>	γ (n) =	3
n =	715	=>	γ (n) =	168	n =	771	=>	γ (n) =	168
n =	716	=>	γ (n) =	20	n =	772	=>	γ (n) =	6
n =	717	=>	γ (n) =	10	n =	773	=>	γ (n) =	10
n =	718	=>	γ (n) =	6	n =	774	=>	γ (n) =	3
n =	719	=>	γ (n) =	6	n =	775	=>	γ (n) =	50
n =	720	=>	γ (n) =	3	n =	776	=>	γ (n) =	10
n =	721	=>	γ (n) =	1176	n =	777	=>	γ (n) =	15
n =	722	=>	γ (n) =	6	n =	778	=>	γ (n) =	20

Table 1: Degrees of cardinal completeness $\gamma(\text{Ł}_n)$

n = 779	=>	$\gamma(n)$ =	6		n = 835	=>	$\gamma(n)$ =	20
n = 780	=>	$\gamma(n)$ =	6		n = 836	=>	$\gamma(n)$ =	6
n = 781	=>	$\gamma(n)$ =	887		n = 837	=>	$\gamma(n)$ =	50
n = 782	=>	$\gamma(n)$ =	6		n = 838	=>	$\gamma(n)$ =	15
n = 783	=>	$\gamma(n)$ =	20		n = 839	=>	$\gamma(n)$ =	6
n = 784	=>	$\gamma(n)$ =	15		n = 840	=>	$\gamma(n)$ =	3
n = 785	=>	$\gamma(n)$ =	56		n = 841	=>	$\gamma(n)$ =	3490
n = 786	=>	$\gamma(n)$ =	6		n = 842	=>	$\gamma(n)$ =	4
n = 787	=>	$\gamma(n)$ =	20		n = 843	=>	$\gamma(n)$ =	6
n = 788	=>	$\gamma(n)$ =	3		n = 844	=>	$\gamma(n)$ =	6
n = 789	=>	$\gamma(n)$ =	10		n = 845	=>	$\gamma(n)$ =	10
n = 790	=>	$\gamma(n)$ =	6		n = 846	=>	$\gamma(n)$ =	10
n = 791	=>	$\gamma(n)$ =	20		n = 847	=>	$\gamma(n)$ =	50
n = 792	=>	$\gamma(n)$ =	6		n = 848	=>	$\gamma(n)$ =	10
n = 793	=>	$\gamma(n)$ =	490		n = 849	=>	$\gamma(n)$ =	21
n = 794	=>	$\gamma(n)$ =	6		n = 850	=>	$\gamma(n)$ =	6
n = 795	=>	$\gamma(n)$ =	6		n = 851	=>	$\gamma(n)$ =	50
n = 796	=>	$\gamma(n)$ =	20		n = 852	=>	$\gamma(n)$ =	6
n = 797	=>	$\gamma(n)$ =	10		n = 853	=>	$\gamma(n)$ =	50
n = 798	=>	$\gamma(n)$ =	3		n = 854	=>	$\gamma(n)$ =	3
n = 799	=>	$\gamma(n)$ =	168		n = 855	=>	$\gamma(n)$ =	20
n = 800	=>	$\gamma(n)$ =	6		n = 856	=>	$\gamma(n)$ =	50
n = 801	=>	$\gamma(n)$ =	84		n = 857	=>	$\gamma(n)$ =	15
n = 802	=>	$\gamma(n)$ =	10		n = 858	=>	$\gamma(n)$ =	3
n = 803	=>	$\gamma(n)$ =	6		n = 859	=>	$\gamma(n)$ =	168
n = 804	=>	$\gamma(n)$ =	6		n = 860	=>	$\gamma(n)$ =	3
n = 805	=>	$\gamma(n)$ =	50		n = 861	=>	$\gamma(n)$ =	50
n = 806	=>	$\gamma(n)$ =	20		n = 862	=>	$\gamma(n)$ =	20
n = 807	=>	$\gamma(n)$ =	20		n = 863	=>	$\gamma(n)$ =	6
n = 808	=>	$\gamma(n)$ =	6		n = 864	=>	$\gamma(n)$ =	3
n = 809	=>	$\gamma(n)$ =	15		n = 865	=>	$\gamma(n)$ =	210
n = 810	=>	$\gamma(n)$ =	3		n = 866	=>	$\gamma(n)$ =	6
n = 811	=>	$\gamma(n)$ =	196		n = 867	=>	$\gamma(n)$ =	6
n = 812	=>	$\gamma(n)$ =	3		n = 868	=>	$\gamma(n)$ =	10
n = 813	=>	$\gamma(n)$ =	50		n = 869	=>	$\gamma(n)$ =	50
n = 814	=>	$\gamma(n)$ =	6		n = 870	=>	$\gamma(n)$ =	6
n = 815	=>	$\gamma(n)$ =	20		n = 871	=>	$\gamma(n)$ =	168
n = 816	=>	$\gamma(n)$ =	6		n = 872	=>	$\gamma(n)$ =	6
n = 817	=>	$\gamma(n)$ =	196		n = 873	=>	$\gamma(n)$ =	15
n = 818	=>	$\gamma(n)$ =	6		n = 874	=>	$\gamma(n)$ =	10
n = 819	=>	$\gamma(n)$ =	6		n = 875	=>	$\gamma(n)$ =	20
n = 820	=>	$\gamma(n)$ =	50		n = 876	=>	$\gamma(n)$ =	15
n = 821	=>	$\gamma(n)$ =	50		n = 877	=>	$\gamma(n)$ =	50
n = 822	=>	$\gamma(n)$ =	3		n = 878	=>	$\gamma(n)$ =	3
n = 823	=>	$\gamma(n)$ =	20		n = 879	=>	$\gamma(n)$ =	6
n = 824	=>	$\gamma(n)$ =	3		n = 880	=>	$\gamma(n)$ =	6
n = 825	=>	$\gamma(n)$ =	15		n = 881	=>	$\gamma(n)$ =	196
n = 826	=>	$\gamma(n)$ =	50		n = 882	=>	$\gamma(n)$ =	3
n = 827	=>	$\gamma(n)$ =	20		n = 883	=>	$\gamma(n)$ =	175
n = 828	=>	$\gamma(n)$ =	3		n = 884	=>	$\gamma(n)$ =	3
n = 829	=>	$\gamma(n)$ =	175		n = 885	=>	$\gamma(n)$ =	50
n = 830	=>	$\gamma(n)$ =	3		n = 886	=>	$\gamma(n)$ =	20
n = 831	=>	$\gamma(n)$ =	20		n = 887	=>	$\gamma(n)$ =	6
n = 832	=>	$\gamma(n)$ =	6		n = 888	=>	$\gamma(n)$ =	3
n = 833	=>	$\gamma(n)$ =	36		n = 889	=>	$\gamma(n)$ =	105
n = 834	=>	$\gamma(n)$ =	10		n = 890	=>	$\gamma(n)$ =	6

Łukasiewicz Logics and Prime Numbers

n = 891	=>	γ (n) =	20	n = 947	=>	γ (n) =	20
n = 892	=>	γ (n) =	21	n = 948	=>	γ (n) =	3
n = 893	=>	γ (n) =	10	n = 949	=>	γ (n) =	50
n = 894	=>	γ (n) =	6	n = 950	=>	γ (n) =	6
n = 895	=>	γ (n) =	20	n = 951	=>	γ (n) =	50
n = 896	=>	γ (n) =	6	n = 952	=>	γ (n) =	6
n = 897	=>	γ (n) =	45	n = 953	=>	γ (n) =	105
n = 898	=>	γ (n) =	20	n = 954	=>	γ (n) =	3
n = 899	=>	γ (n) =	6	n = 955	=>	γ (n) =	50
n = 900	=>	γ (n) =	6	n = 956	=>	γ (n) =	6
n = 901	=>	γ (n) =	980	n = 957	=>	γ (n) =	10
n = 902	=>	γ (n) =	6	n = 958	=>	γ (n) =	20
n = 903	=>	γ (n) =	20	n = 959	=>	γ (n) =	6
n = 904	=>	γ (n) =	20	n = 960	=>	γ (n) =	6
n = 905	=>	γ (n) =	15	n = 961	=>	γ (n) =	540
n = 906	=>	γ (n) =	6	n = 962	=>	γ (n) =	4
n = 907	=>	γ (n) =	20	n = 963	=>	γ (n) =	20
n = 908	=>	γ (n) =	3	n = 964	=>	γ (n) =	10
n = 909	=>	γ (n) =	10	n = 965	=>	γ (n) =	10
n = 910	=>	γ (n) =	10	n = 966	=>	γ (n) =	6
n = 911	=>	γ (n) =	168	n = 967	=>	γ (n) =	168
n = 912	=>	γ (n) =	3	n = 968	=>	γ (n) =	3
n = 913	=>	γ (n) =	196	n = 969	=>	γ (n) =	35
n = 914	=>	γ (n) =	6	n = 970	=>	γ (n) =	20
n = 915	=>	γ (n) =	6	n = 971	=>	γ (n) =	20
n = 916	=>	γ (n) =	20	n = 972	=>	γ (n) =	3
n = 917	=>	γ (n) =	10	n = 973	=>	γ (n) =	84
n = 918	=>	γ (n) =	6	n = 974	=>	γ (n) =	6
n = 919	=>	γ (n) =	105	n = 975	=>	γ (n) =	6
n = 920	=>	γ (n) =	3	n = 976	=>	γ (n) =	50
n = 921	=>	γ (n) =	105	n = 977	=>	γ (n) =	21
n = 922	=>	γ (n) =	6	n = 978	=>	γ (n) =	3
n = 923	=>	γ (n) =	6	n = 979	=>	γ (n) =	20
n = 924	=>	γ (n) =	6	n = 980	=>	γ (n) =	6
n = 925	=>	γ (n) =	887	n = 981	=>	γ (n) =	175
n = 926	=>	γ (n) =	10	n = 982	=>	γ (n) =	10
n = 927	=>	γ (n) =	6	n = 983	=>	γ (n) =	6
n = 928	=>	γ (n) =	10	n = 984	=>	γ (n) =	3
n = 929	=>	γ (n) =	28	n = 985	=>	γ (n) =	105
n = 930	=>	γ (n) =	3	n = 986	=>	γ (n) =	6
n = 931	=>	γ (n) =	168	n = 987	=>	γ (n) =	20
n = 932	=>	γ (n) =	10	n = 988	=>	γ (n) =	20
n = 933	=>	γ (n) =	10	n = 989	=>	γ (n) =	50
n = 934	=>	γ (n) =	6	n = 990	=>	γ (n) =	6
n = 935	=>	γ (n) =	6	n = 991	=>	γ (n) =	887
n = 936	=>	γ (n) =	20	n = 992	=>	γ (n) =	3
n = 937	=>	γ (n) =	490	n = 993	=>	γ (n) =	28
n = 938	=>	γ (n) =	3	n = 994	=>	γ (n) =	6
n = 939	=>	γ (n) =	20	n = 995	=>	γ (n) =	20
n = 940	=>	γ (n) =	6	n = 996	=>	γ (n) =	6
n = 941	=>	γ (n) =	50	n = 997	=>	γ (n) =	50
n = 942	=>	γ (n) =	3	n = 998	=>	γ (n) =	3
n = 943	=>	γ (n) =	20	n = 999	=>	γ (n) =	6
n = 944	=>	γ (n) =	6	n = 1000	=>	γ (n) =	15
n = 945	=>	γ (n) =	21				
n = 946	=>	γ (n) =	105				

Table 2
Cardinality of rooted trees T_p and CRT
(see section V.3 *and* V.5)*

p	T_p	CRT		p	T_p	CRT
				239	2	1
2	2	1		241	186	19
3	3	1		251	2	1
5	4	1		257	10	1
7	9	2		263	2	1
11	2	1		269	2	1
13	31	5		271	2	1
17	6	1		277	18	3
19	4	1		281	8	1
23	2	1		283	2	1
29	2	1		293	2	1
31	2	1		307	2	1
37	11	2		311	2	1
41	24	4		313	96	11
43	41	6		317	2	1
47	2	1		331	2	1
53	2	1		337	18	1
59	2	2		347	2	1
61	57	9		349	6	1
67	2	1		353	15	2
71	2	1		359	2	1
73	58	7		367	2	1
79	2	1		373	2	1
83	2	1		379	2	1
89	6	1		383	2	1
97	17	1		389	2	1
101	4	1		397	44	6
103	2	1		401	34	3
107	2	1		409	6	1
109	39	5		419	2	1
113	67	6		421	16	2
127	2	1		431	2	1
131	2	1		433	105	11
137	2	1		439	2	1
139	2	1		443	2	1
149	2	1		449	60	9
151	2	1		457	5	1
157	25	4		461	4	1
163	4	1		463	2	1
167	2	1		467	2	1
173	2	1		479	2	1
179	2	1		487	4	1
181	158	16		491	2	1
191	2	1		499	2	1
193	61	9		503	2	1
197	2	2		509	2	1
199	2	1		521	8	1
211	2	1		523	2	1
223	2	1		541	84	9
227	2	1		547	2	1
229	2	1		557	2	1
233	54	7		563	2	1

569	2	1		883	2	1
571	2	1		887	2	1
577	54	3		907	2	1
587	2	1		911	2	1
593	9	2		919	2	1
599	2	1		929	12	1
601	24	2		937	39	4
607	2	1		941	2	1
613	292	30		947	2	1
617	4	1		953	6	1
619	2	1		967	2	1
631	2	1		971	2	1
641	30	2		977	2	1
643	2	1		983	2	1
647	2	1		991	2	1
653	2	1		997	20	3
659	2	1		1009	142	16
661	32	4		1013	57	5
673	70	6		1019	2	1
677	2	1		1021	4	1
683	2	1		1031	2	1
691	2	1		1033	8	2
701	10	2		1039	2	1
709	2	1		1049	6	1
719	2	1		1051	2	1
727	2	1		1061	19	3
733	5	1		1063	2	1
739	2	1		1069	6	1
743	2	1		1087	2	1
751	2	1		1091	2	1
757	141	14		1093	281	28
761	6	1		1097	2	1
769	58	4		1103	2	1
773	2	1		1109	2	1
787	2	1		1117	2	1
797	2	1		1123	2	1
809	2	1		1129	6	1
811	2	1		1151	2	1
821	13	3		1153	290	31
823	2	1		1163	2	1
827	2	1		1171	2	1
829	32	4		1181	2	1
839	2	1		1187	2	1
853	2	1		1193	2	1
857	2	1		1201	72	7
859	2	1		1213	11	3
863	2	1		1217	2	1
877	8	2		1223	2	1
881	132	14				

Table 2 : Cardinality of rooted trees T_p and CRT

1229	2	1		1583	2	1
1231	2	1		1597	2	1
1237	20	2		1601	39	4
1249	35	2		1607	2	1
1259	2	1		1609	14	3
1277	8	2		1613	2	1
1279	2	1		1619	2	1
1283	2	1		1621	224	22
1289	4	1		1627	2	1
1291	2	1		1637	2	1
1297	463	40		1657	48	5
1301	96	6		1663	2	1
1303	2	1		1667	2	1
1307	2	1		1669	2	1
1319	2	1		1693	37	6
1321	203	18		1697	8	1
1327	2	1		1699	2	1
1361	4	1		1709	2	1
1367	2	1		1721	8	1
1373	2	1		1723	2	1
1381	54	4		1733	2	1
1399	2	1		1741	4	1
1409	103	11		1747	2	1
1423	2	1		1753	33	5
1427	2	1		1759	2	1
1429	10	2		1777	20	3
1433	6	1		1783	2	1
1439	2	1		1787	2	1
1447	2	1		1789	2	1
1451	2	1		1801	24	1
1453	71	11		1811	2	1
1459	4	1		1823	2	1
1471	2	1		1831	2	1
1481	4	1		1847	2	1
1483	2	1		1861	24	3
1487	2	1		1867	2	1
1489	10	2		1871	2	1
1493	2	1		1873	222	17
1499	2	1		1877	2	1
1511	2	1		1879	2	1
1523	2	1		1889	2	1
1531	2	1		1901	142	19
1543	2	1		1907	2	1
1549	2	1		1913	6	1
1553	63	8		1931	2	1
1559	2	1		1933	23	3
1567	2	1		1949	2	1
1571	2	1		1951	2	1
1579	2	1		1973	2	1
				1979	2	1
				1987	2	1

1993	15	1	2371	2	1
1997	2	1	2377	52	8
1999	2	1	2381	191	14
2003	2	1	2383	2	1
2011	2	1	2389	2	1
2017	105	7	2393	8	1
2027	2	1	2399	2	1
2029	12	3	2411	2	1
2039	2	1	2417	2	1
2053	13	2	2423	2	1
2063	2	1	2437	6	1
2069	2	1	2441	2	1
2081	27	3	2447	2	1
2083	2	1	2459	2	1
2087	2	1	2467	2	1
2089	24	3	2473	9	1
2099	2	1	2477	2	1
2111	2	1	2503	2	1
2113	72	6	2521	70	6
2129	2	1	2531	2	1
2131	2	1	2539	2	1
2137	11	1	2543	2	1
2141	2	1	2549	2	1
2143	2	1	2551	2	1
2153	2	1	2557	5	1
2161	213	22	2579	2	1
2179	2	1	2591	2	1
2203	2	1	2593	128	11
2207	2	1	2609	49	7
2213	2	1	2617	2	1
2221	4	1	2621	7	2
2237	2	1	2633	6	1
2239	2	1	2647	2	1
2243	2	1	2657	8	1
2251	2	1	2659	2	1
2267	2	1	2663	2	1
2269	52	9	2671	2	1
2273	9	2	2677	2	1
2281	14	1	2683	2	1
2287	2	1	2687	2	1
2293	8	2	2689	64	3
2297	4	1	2693	2	1
2309	2	1	2699	2	1
2311	2	1	2707	2	1
2333	12	3	2711	2	1
2339	2	1	2713	8	1
2341	72	8	2719	2	1
2347	2	1	2729	6	1
2351	2	1	2731	2	1
2357	2	1	2741	2	1

Table 2 : Cardinality of rooted trees T_p and CRT

2749	2	1		3187	2	1
2753	39	8		3191	2	1
2767	2	1		3203	2	1
2777	2	1		3209	2	1
2789	2	1		3217	11	1
2791	2	1		3221	11	2
2797	12	2		3229	2	1
2801	97	13		3251	2	1
2803	2	1		3253	8	2
2819	2	1		3257	4	1
2833	10	2		3259	2	1
2837	2	1		3271	2	1
2843	2	1		3299	2	1
2851	2	1		3301	16	2
2857	13	1		3307	2	1
2861	72	10		3313	2125	190
2879	2	1		3319	2	1
2887	2	1		3323	2	1
2897	2	1		3329	10	1
2903	2	1		3331	2	1
2909	2	1		3343	2	1
2917	131	14		3347	2	1
2927	2	1		3359	2	1
2939	2	1		3361	422	28
2953	14	1		3371	2	1
2957	2	1		3373	6	1
2963	2	1		3389	2	1
2969	8	1		3391	2	1
2971	2	1		3407	2	1
2999	2	1		3413	2	1
3001	77	8		3433	15	1
3011	2	1		3449	6	1
3019	2	1		3457	209	15
3023	2	1		3461	22	4
3037	18	2		3463	2	1
3041	12	1		3467	2	1
3049	6	1		3469	2	1
3061	18	3		3491	2	1
3067	2	1		3499	2	1
3079	2	1		3511	2	1
3083	2	1		3517	9	1
3089	21	2		3527	2	1
3109	2	1		3529	26	3
3119	2	1		3533	2	1
3121	405	36		3539	2	1
3137	26	5		3541	2	1
3163	2	1		3547	2	1
3167	2	1		3557	2	1
3169	56	2		3559	2	1
3181	4	1		3571	2	1

3581	8	2		4001	39	4
3583	2	1		4003	2	1
3593	2	1		4007	2	1
3607	2	1		4013	2	1
3613	5	1		4019	2	1
3617	8	1		4021	11	3
3623	2	1		4027	2	1
3631	2	1		4049	17	1
3637	27	5		4051	2	1
3643	2	1		4057	14	1
3659	2	1		4073	6	1
3671	2	1		4079	2	1
3673	24	1		4091	2	1
3677	2	1		4093	8	2
3691	2	1		4099	2	1
3697	52	5		4111	2	1
3701	2	1		4127	2	1
3709	6	1		4129	16	1
3719	2	1		4133	2	1
3727	2	1		4139	2	1
3733	39	3		4153	12	1
3739	2	1		4157	2	1
3761	76	6		4159	2	1
3767	2	1		4177	425	32
3769	2	1		4201	36	1
3779	2	1		4211	2	1
3793	24	3		4217	2	1
3797	2	1		4219	2	1
3803	2	1		4229	2	1
3821	6	2		4231	2	1
3823	2	1		4241	30	3
3833	2	1		4243	2	1
3847	2	1		4253	2	1
3851	2	1		4259	2	1
3853	192	21		4261	5	1
3863	2	1		4271	2	1
3877	6	1		4273	51	4
3881	8	1		4283	2	1
3889	542	38		4289	11	2
3907	2	1		4297	8	1
3911	2	1		4327	2	1
3917	7	2		4337	2	1
3919	2	1		4339	2	1
3923	2	1		4349	2	1
3929	6	1		4357	134	15
3931	2	1		4363	2	1
3943	2	1		4373	2	1
3947	2	1		4391	2	1
3967	2	1		4397	2	1
3989	2	1		4409	6	1

Table 2 : Cardinality of rooted trees T_p and CRT

4421	36	6	4861	834	70
4423	11	2	4871	2	1
4441	12	1	4877	8	2
4447	2	1	4889	6	1
4451	2	1	4903	2	1
4457	2	1	4909	2	1
4463	2	1	4919	2	1
4481	54	4	4931	2	1
4483	2	1	4933	161	7
4493	2	1	4937	2	1
4507	2	1	4943	2	1
4513	12	1	4951	2	1
4517	2	1	4957	6	1
4519	2	1	4967	2	1
4523	2	1	4969	33	3
4547	2	1	4973	10	3
4549	2	1	4987	2	1
4561	61	9	4993	53	2
4567	2	1	4999	2	1
4583	2	1	5003	2	1
4591	2	1	5009	2	1
4597	2	1	5011	2	1
4603	2	1	5021	7	2
4621	164	18	5023	2	1
4637	2	1	5039	2	1
4639	2	1	5051	2	1
4643	2	1	5059	2	1
4649	8	1	5077	230	18
4651	2	1	5081	2	1
4657	14	2	5087	2	1
4663	2	1	5099	2	1
4673	180	16	5101	5	1
4679	2	1	5107	2	1
4691	2	1	5113	13	1
4703	2	1	5119	2	1
4721	50	7	5147	2	1
4723	2	1	5153	74	9
4729	2	1	5167	2	1
4733	2	1	5171	2	1
4751	2	1	5179	2	1
4759	2	1	5189	2	1
4783	2	1	5197	2	1
4787	2	1	5209	6	1
4789	2	1	5227	2	1
4793	2	1	5231	2	1
4799	2	1	5233	5	1
4801	268	24	5237	7	2
4813	2	1	5261	2	1
4817	12	2	5273	6	1
4831	2	1	5279	2	1

Łukasiewicz Logics and Prime Numbers

5281	177	14		5701	32	4
5297	2	1		5711	2	1
5303	2	1		5717	2	1
5309	2	1		5737	8	1
5323	2	1		5741	61	12
5333	2	1		5743	2	1
5347	2	1		5749	2	1
5351	2	1		5779	2	1
5381	2	1		5783	2	1
5387	2	1		5791	2	1
5393	2	1		5801	10	1
5399	2	1		5807	2	1
5407	2	1		5813	2	1
5413	14	3		5821	6	1
5417	2	1		5827	2	1
5419	2	1		5839	2	1
5431	2	1		5843	2	1
5437	14	3		5849	2	1
5441	639	53		5851	2	1
5443	2	1		5857	20	2
5449	2	1		5861	7	2
5471	2	1		5867	2	1
5477	2	1		5869	2	1
5479	2	1		5879	2	1
5483	2	1		5881	27	1
5501	7	2		5897	4	1
5503	2	1		5903	2	1
5507	2	1		5923	2	1
5519	2	1		5927	2	1
5521	738	66		5939	2	1
5527	2	1		5953	12	1
5531	2	1		5981	41	8
5557	2	1		5987	2	1
5563	2	1		6007	2	1
5569	90	8		6011	2	1
5573	2	1		6029	2	1
5581	27	5		6037	5	1
5591	2	1		6043	2	1
5623	2	1		6047	2	1
5639	2	1		6053	2	1
5641	11	1		6067	2	1
5647	2	1		6073	42	5
5651	2	1		6079	2	1
5653	2	1		6089	6	1
5657	2	1		6091	2	1
5659	2	1		6101	2	1
5669	2	1		6113	8	1
5683	2	1		6121	47	3
5689	10	1		6131	2	1
5693	2	1		6133	72	11

Table 2 : Cardinality of rooted trees T_p and CRT

6143	2	1		6577	17	3
6151	2	1		6581	12	3
6163	90	6		6599	2	1
6173	2	1		6607	2	1
6197	2	1		6619	2	1
6199	2	1		6637	5	1
6203	2	1		6653	2	1
6211	2	1		6659	2	1
6217	11	1		6661	826	59
6221	2	1		6673	23	4
6229	23	2		6679	2	1
6247	2	1		6689	8	1
6257	4	1		6691	2	1
6263	2	1		6701	2	1
6269	2	1		6703	2	1
6271	2	1		6709	2	1
6277	2	1		6719	2	1
6287	2	1		6733	293	26
6299	2	1		6737	2	1
6301	23	3		6761	6	1
6311	2	1		6763	2	1
6317	2	1		6779	2	1
6323	2	1		6781	10	2
6329	8	1		6791	2	1
6337	112	6		6793	6	1
6343	2	1		6803	2	1
6353	2	1		6823	2	1
6359	2	1		6827	2	1
6361	21	2		6829	2	1
6367	2	1		6833	29	2
6373	48	5		6841	62	7
6379	2	1		6857	2	1
6389	2	1		6863	2	1
6397	4	1		6869	2	1
6421	31	5		6871	2	1
6427	2	1		6883	2	1
6449	1325	108		6899	2	1
6451	2	1		6907	2	1
6469	23	3		6911	2	1
6473	6	1		6917	2	1
6481	1122	87		6947	2	1
6491	2	1		6949	2	1
6521	4	1		6959	2	1
6529	54	6		6961	469	39
6547	2	1		6967	2	1
6551	2	1		6971	2	1
6553	100	6		6977	2	1
6563	2	1		6983	2	1
6569	2	1		6991	2	1
6571	2	1		6997	89	12

Łukasiewicz Logics and Prime Numbers

7001	10	1		7523	2	1
7013	2	1		7529	2	1
7019	2	1		7537	5	1
7027	2	1		7541	255	25
7039	2	1		7547	2	1
7043	2	1		7549	6	1
7057	51	3		7559	2	1
7069	2	1		7561	109	5
7079	2	1		7573	2	1
7103	2	1		7577	2	1
7109	2	1		7583	2	1
7121	11	1		7589	2	1
7127	2	1		7591	2	1
7129	151	15		7603	2	1
7151	2	1		7607	2	1
7159	2	1		7621	2	1
7177	17	1		7639	2	1
7187	2	1		7643	2	1
7193	2	1		7649	15	2
7207	2	1		7669	20	4
7211	2	1		7673	2	1
7213	5	1		7681	117	4
7219	2	1		7687	2	1
7229	2	1		7691	2	1
7237	2	1		7699	2	1
7243	2	1		7703	2	1
7247	2	1		7717	2	1
7253	2	1		7723	2	1
7283	2	1		7727	2	1
7297	22	2		7741	43	7
7307	2	1		7753	11	1
7309	4	1		7757	2	1
7321	16	1		7759	2	1
7331	2	1		7789	2	1
7333	22	4		7793	92	10
7349	2	1		7817	2	1
7351	2	1		7823	2	1
7369	12	3		7829	2	1
7393	53	2		7841	25	1
7411	2	1		7853	2	1
7417	15	1		7867	2	1
7433	2	1		7873	22	2
7451	2	1		7877	26	3
7457	8	1		7879	2	1
7459	2	1		7883	2	1
7477	19	3		7901	2	1
7481	6	1		7907	2	1
7487	2	1		7919	2	1
7489	112	9				
7499	2	1				
7507	2	1				
7517	2	1				

Table 3
Values of function *i(p)*: classes of prime numbers
(see section VI.5)

i(2)=0	i(193) = 4	i(449) = 8
i(3) = 0	i(197) = 4	i(457) = 8
i(5) = 0	i(199) = 4	i(461) = 8
i(7) = 0	i(211) = 4	i(463) = 8
i(11) = 0	i(223) = 8	i(467) = 7
i(13) = 0	i(227) = 8	i(479) = 6
i(17) = 1	i(229) = 8	i(487) = 6
i(19) = 1	i(233) = 8	i(491) = 6
i(23) = 2	i(239) = 6	i(499) = 6
i(29) = 2	i(241) = 5	i(503) = 6
i(31) = 2	i(251) = 4	i(509) = 5
i(37) = 3	i(257) = 4	i(521) = 5
i(41) = 2	i(263) = 4	i(523) = 4
i(43) = 2	i(269) = 4	i(541) = 4
i(47) = 3	i(271) = 4	i(547) = 4
i(53) = 2	i(277) = 6	i(557) = 6
i(59) = 2	i(281) = 4	i(563) = 6
i(61) = 2	i(283) = 5	i(569) = 5
i(67) = 4	i(293) = 4	i(571) = 5
i(71) = 4	i(307) = 5	i(577) = 5
i(73) = 4	i(311) = 4	i(587) = 4
i(79) = 4	i(313) = 5	i(593) = 6
i(83) = 4	i(317) = 5	i(599) = 5
i(89) = 4	i(331) = 4	i(601) = 6
i(97) = 4	i(337) = 4	i(607) = 5
i(101) = 4	i(347) = 4	i(613) = 5
i(103) = 4	i(349) = 4	i(617) = 6
i(107) = 4	i(353) = 5	i(619) = 5
i(109) = 3	i(359) = 6	i(631) = 5
i(113) = 4	i(367) = 5	i(641) = 8
i(127) = 4	i(373) = 4	i(643) = 8
i(131) = 4	i(379) = 6	i(647) = 8
i(137) = 4	i(383) = 4	i(653) = 8
i(139) = 5	i(389) = 4	i(659) = 8
i(149) = 4	i(397) = 4	i(661) = 8
i(151) = 4	i(401) = 4	i(673) = 8
i(157) = 4	i(409) = 4	i(677) = 8
i(163) = 4	i(419) = 4	i(683) = 8
i(167) = 4	i(421) = 4	i(691) = 8
i(173) = 4	i(431) = 8	i(701) = 8
i(179) = 4	i(433) = 8	i(709) = 9
i(181) = 4	i(439) = 8	i(719) = 6
i(191) = 4	i(443) = 8	i(727) = 5

i(733) = 6	i(1061) = 8	i(1429) = 8
i(739) = 6	i(1063) = 8	i(1433) = 8
i(743) = 6	i(1069) = 8	i(1439) = 7
i(751) = 6	i(1087) = 8	i(1447) = 7
i(757) = 5	i(1091) = 8	i(1451) = 8
i(761) = 6	i(1093) = 8	i(1453) = 8
i(769) = 6	i(1097) = 8	i(1459) = 6
i(773) = 6	i(1103) = 8	i(1471) = 8
i(787) = 6	i(1109) = 8	i(1481) = 8
i(797) = 6	i(1117) = 8	i(1483) = 8
i(809) = 6	i(1123) = 8	i(1487) = 8
i(811) = 6	i(1129) = 8	i(1489) = 8
i(821) = 6	i(1151) = 8	i(1493) = 8
i(823) = 7	i(1153) = 8	i(1499) = 8
i(827) = 6	i(1163) = 8	i(1511) = 8
i(829) = 6	i(1171) = 7	i(1523) = 8
i(839) = 6	i(1181) = 7	i(1531) = 8
i(853) = 8	i(1187) = 6	i(1543) = 8
i(857) = 8	i(1193) = 7	i(1549) = 8
i(859) = 8	i(1201) = 6	i(1553) = 8
i(863) = 8	i(1213) = 6	i(1559) = 8
i(877) = 8	i(1217) = 7	i(1567) = 8
i(881) = 8	i(1223) = 6	i(1571) = 8
i(883) = 8	i(1229) = 6	i(1579) = 8
i(887) = 8	i(1231) = 6	i(1583) = 8
i(907) = 8	i(1237) = 5	i(1597) = 8
i(911) = 8	i(1249) = 6	i(1601) = 8
i(919) = 8	i(1259) = 6	i(1607) = 8
i(929) = 8	i(1277) = 8	i(1609) = 9
i(937) = 7	i(1279) = 8	i(1613) = 8
i(941) = 7	i(1283) = 8	i(1619) = 8
i(947) = 6	i(1289) = 8	i(1621) = 8
i(953) = 6	i(1291) = 8	i(1627) = 8
i(967) = 6	i(1297) = 8	i(1637) = 7
i(971) = 7	i(1301) = 8	i(1657) = 8
i(977) = 6	i(1303) = 8	i(1663) = 8
i(983) = 6	i(1307) = 8	i(1667) = 7
i(991) = 6	i(1319) = 8	i(1669) = 7
i(997) = 6	i(1321) = 8	i(1693) = 8
i(1009) = 6	i(1327) = 8	i(1697) = 8
i(1013) = 6	i(1361) = 8	i(1699) = 8
i(1019) = 6	i(1367) = 8	i(1709) = 8
i(1021) = 6	i(1373) = 8	i(1721) = 8
i(1031) = 6	i(1381) = 8	i(1723) = 8
i(1033) = 7	i(1399) = 8	i(1733) = 8
i(1039) = 6	i(1409) = 8	i(1741) = 8
i(1049) = 6	i(1423) = 8	i(1747) = 8
i(1051) = 6	i(1427) = 9	i(1753) = 8

Table 3 : Values of function $i(p)$: classes of prime numbers

$i(1759) = 8$	$i(2137) = 9$	$i(2531) = 9$
$i(1777) = 8$	$i(2141) = 10$	$i(2539) = 9$
$i(1783) = 8$	$i(2143) = 8$	$i(2543) = 8$
$i(1787) = 8$	$i(2153) = 8$	$i(2549) = 8$
$i(1789) = 8$	$i(2161) = 8$	$i(2551) = 8$
$i(1801) = 8$	$i(2179) = 8$	$i(2557) = 8$
$i(1811) = 8$	$i(2203) = 8$	$i(2579) = 8$
$i(1823) = 8$	$i(2207) = 8$	$i(2591) = 8$
$i(1831) = 8$	$i(2213) = 8$	$i(2593) = 8$
$i(1847) = 9$	$i(2221) = 9$	$i(2609) = 8$
$i(1861) = 8$	$i(2237) = 8$	$i(2617) = 8$
$i(1867) = 8$	$i(2239) = 8$	$i(2621) = 8$
$i(1871) = 7$	$i(2243) = 8$	$i(2633) = 8$
$i(1873) = 8$	$i(2251) = 8$	$i(2647) = 8$
$i(1877) = 7$	$i(2267) = 9$	$i(2657) = 8$
$i(1879) = 7$	$i(2269) = 8$	$i(2659) = 8$
$i(1889) = 7$	$i(2273) = 8$	$i(2663) = 8$
$i(1901) = 8$	$i(2281) = 8$	$i(2671) = 8$
$i(1907) = 8$	$i(2287) = 8$	$i(2677) = 8$
$i(1913) = 8$	$i(2293) = 8$	$i(2683) = 8$
$i(1931) = 8$	$i(2297) = 8$	$i(2687) = 8$
$i(1933) = 8$	$i(2309) = 8$	$i(2689) = 8$
$i(1949) = 8$	$i(2311) = 8$	$i(2693) = 8$
$i(1951) = 8$	$i(2333) = 10$	$i(2699) = 8$
$i(1973) = 8$	$i(2339) = 10$	$i(2707) = 8$
$i(1979) = 8$	$i(2341) = 10$	$i(2711) = 8$
$i(1987) = 8$	$i(2347) = 10$	$i(2713) = 8$
$i(1993) = 10$	$i(2351) = 10$	$i(2719) = 8$
$i(1997) = 9$	$i(2357) = 10$	$i(2729) = 8$
$i(1999) = 8$	$i(2371) = 10$	$i(2731) = 8$
$i(2003) = 8$	$i(2377) = 10$	$i(2741) = 10$
$i(2011) = 9$	$i(2381) = 10$	$i(2749) = 8$
$i(2017) = 8$	$i(2383) = 10$	$i(2753) = 8$
$i(2027) = 9$	$i(2389) = 10$	$i(2767) = 8$
$i(2029) = 9$	$i(2393) = 10$	$i(2777) = 8$
$i(2039) = 8$	$i(2399) = 11$	$i(2789) = 8$
$i(2053) = 8$	$i(2411) = 10$	$i(2791) = 9$
$i(2063) = 8$	$i(2417) = 10$	$i(2797) = 8$
$i(2069) = 8$	$i(2423) = 10$	$i(2801) = 8$
$i(2081) = 8$	$i(2437) = 10$	$i(2803) = 8$
$i(2083) = 8$	$i(2441) = 10$	$i(2819) = 8$
$i(2087) = 8$	$i(2447) = 10$	$i(2833) = 8$
$i(2089) = 8$	$i(2459) = 10$	$i(2837) = 8$
$i(2099) = 8$	$i(2467) = 10$	$i(2843) = 9$
$i(2111) = 8$	$i(2473) = 10$	$i(2851) = 8$
$i(2113) = 8$	$i(2477) = 10$	$i(2857) = 10$
$i(2129) = 8$	$i(2503) = 10$	$i(2861) = 11$
$i(2131) = 8$	$i(2521) = 9$	$i(2879) = 8$

i(2887) = 8	i(3313) = 8	i(3677) = 8
i(2897) = 8	i(3319) = 8	i(3691) = 10
i(2903) = 8	i(3323) = 10	i(3697) = 8
i(2909) = 8	i(3329) = 8	i(3701) = 9
i(2917) = 8	i(3331) = 8	i(3709) = 8
i(2927) = 9	i(3343) = 8	i(3719) = 8
i(2939) = 8	i(3347) = 8	i(3727) = 8
i(2953) = 9	i(3359) = 9	i(3733) = 8
i(2957) = 8	i(3361) = 8	i(3739) = 8
i(2963) = 8	i(3371) = 8	i(3761) = 8
i(2969) = 8	i(3373) = 8	i(3767) = 8
i(2971) = 8	i(3389) = 8	i(3769) = 8
i(2999) = 8	i(3391) = 8	i(3779) = 11
i(3001) = 8	i(3407) = 8	i(3793) = 8
i(3011) = 8	i(3413) = 10	i(3797) = 9
i(3019) = 8	i(3433) = 8	i(3803) = 8
i(3023) = 9	i(3449) = 8	i(3821) = 9
i(3037) = 8	i(3457) = 9	i(3823) = 9
i(3041) = 8	i(3461) = 8	i(3833) = 8
i(3049) = 8	i(3463) = 8	i(3847) = 8
i(3061) = 8	i(3467) = 8	i(3851) = 10
i(3067) = 9	i(3469) = 8	i(3853) = 8
i(3079) = 8	i(3491) = 8	i(3863) = 8
i(3083) = 8	i(3499) = 8	i(3877) = 9
i(3089) = 8	i(3511) = 8	i(3881) = 8
i(3109) = 8	i(3517) = 8	i(3889) = 8
i(3119) = 8	i(3527) = 9	i(3907) = 8
i(3121) = 8	i(3529) = 8	i(3911) = 11
i(3137) = 8	i(3533) = 8	i(3917) = 10
i(3163) = 10	i(3539) = 8	i(3919) = 8
i(3167) = 8	i(3541) = 8	i(3923) = 8
i(3169) = 8	i(3547) = 8	i(3929) = 8
i(3181) = 8	i(3557) = 10	i(3931) = 8
i(3187) = 10	i(3559) = 9	i(3943) = 8
i(3191) = 8	i(3571) = 9	i(3947) = 9
i(3203) = 8	i(3581) = 10	i(3967) = 8
i(3209) = 8	i(3583) = 10	i(3989) = 8
i(3217) = 8	i(3593) = 8	i(4001) = 10
i(3221) = 8	i(3607) = 8	i(4003) = 10
i(3229) = 9	i(3613) = 8	i(4007) = 10
i(3251) = 8	i(3617) = 8	i(4013) = 8
i(3253) = 9	i(3623) = 8	i(4019) = 10
i(3257) = 8	i(3631) = 9	i(4021) = 8
i(3259) = 8	i(3637) = 8	i(4027) = 9
i(3271) = 8	i(3643) = 10	i(4049) = 8
i(3299) = 8	i(3659) = 8	i(4051) = 10
i(3301) = 8	i(3671) = 10	i(4057) = 8
i(3307) = 8	i(3673) = 8	i(4073) = 8

Table 3 : Values of function $i(p)$: classes of prime numbers

i(4079) = 9	i(4493) = 10	i(4931) = 10
i(4091) = 9	i(4507) = 9	i(4933) = 10
i(4093) = 8	i(4513) = 9	i(4937) = 10
i(4099) = 9	i(4517) = 8	i(4943) = 10
i(4111) = 8	i(4519) = 10	i(4951) = 10
i(4127) = 9	i(4523) = 9	i(4957) = 10
i(4129) = 9	i(4547) = 9	i(4967) = 10
i(4133) = 10	i(4549) = 8	i(4969) = 10
i(4139) = 9	i(4561) = 8	i(4973) = 10
i(4153) = 8	i(4567) = 8	i(4987) = 10
i(4157) = 10	i(4583) = 8	i(4993) = 10
i(4159) = 10	i(4591) = 8	i(4999) = 10
i(4177) = 8	i(4597) = 8	i(5003) = 10
i(4201) = 8	i(4603) = 10	i(5009) = 10
i(4211) = 10	i(4621) = 8	i(5011) = 10
i(4217) = 8	i(4637) = 12	i(5021) = 10
i(4219) = 8	i(4639) = 12	i(5023) = 10
i(4229) = 8	i(4643) = 12	i(5039) = 10
i(4231) = 9	i(4649) = 13	i(5051) = 12
i(4241) = 8	i(4651) = 12	i(5059) = 9
i(4243) = 8	i(4657) = 11	i(5077) = 10
i(4253) = 8	i(4663) = 10	i(5081) = 9
i(4259) = 8	i(4673) = 10	i(5087) = 8
i(4261) = 9	i(4679) = 10	i(5099) = 8
i(4271) = 8	i(4691) = 10	i(5101) = 9
i(4273) = 9	i(4703) = 10	i(5107) = 8
i(4283) = 9	i(4721) = 10	i(5113) = 9
i(4289) = 10	i(4723) = 10	i(5119) = 10
i(4297) = 10	i(4729) = 10	i(5147) = 8
i(4327) = 8	i(4733) = 10	i(5153) = 10
i(4337) = 8	i(4751) = 10	i(5167) = 8
i(4339) = 10	i(4759) = 10	i(5171) = 9
i(4349) = 8	i(4783) = 10	i(5179) = 8
i(4357) = 8	i(4787) = 10	i(5189) = 8
i(4363) = 8	i(4789) = 10	i(5197) = 10
i(4373) = 8	i(4793) = 10	i(5209) = 9
i(4391) = 8	i(4799) = 10	i(5227) = 8
i(4397) = 9	i(4801) = 10	i(5231) = 9
i(4409) = 10	i(4813) = 10	i(5233) = 10
i(4421) = 8	i(4817) = 10	i(5237) = 8
i(4423) = 10	i(4831) = 10	i(5261) = 10
i(4441) = 8	i(4861) = 10	i(5273) = 10
i(4447) = 10	i(4871) = 10	i(5279) = 9
i(4451) = 10	i(4877) = 10	i(5281) = 9
i(4457) = 10	i(4889) = 10	i(5297) = 8
i(4463) = 10	i(4903) = 10	i(5303) = 10
i(4481) = 8	i(4909) = 10	i(5309) = 9
i(4483) = 8	i(4919) = 10	i(5323) = 12

i(5333) = 8	i(5737) = 9	i(6151) = 10
i(5347) = 8	i(5741) = 9	i(6163) = 10
i(5351) = 8	i(5743) = 10	i(6173) = 10
i(5381) = 9	i(5749) = 8	i(6197) = 10
i(5387) = 9	i(5779) = 10	i(6199) = 10
i(5393) = 8	i(5783) = 11	i(6203) = 10
i(5399) = 8	i(5791) = 8	i(6211) = 10
i(5407) = 8	i(5801) = 10	i(6217) = 10
i(5413) = 8	i(5807) = 10	i(6221) = 10
i(5417) = 9	i(5813) = 10	i(6229) = 10
i(5419) = 8	i(5821) = 10	i(6247) = 10
i(5431) = 8	i(5827) = 8	i(6257) = 10
i(5437) = 8	i(5839) = 8	i(6263) = 10
i(5441) = 8	i(5843) = 10	i(6269) = 10
i(5443) = 9	i(5849) = 8	i(6271) = 10
i(5449) = 8	i(5851) = 10	i(6277) = 12
i(5471) = 12	i(5857) = 8	i(6287) = 11
i(5477) = 12	i(5861) = 9	i(6299) = 10
i(5479) = 10	i(5867) = 10	i(6301) = 10
i(5483) = 10	i(5869) = 9	i(6311) = 12
i(5501) = 8	i(5879) = 8	i(6317) = 10
i(5503) = 9	i(5881) = 9	i(6323) = 10
i(5507) = 8	i(5897) = 10	i(6329) = 10
i(5519) = 8	i(5903) = 8	i(6337) = 10
i(5521) = 8	i(5923) = 9	i(6343) = 10
i(5527) = 9	i(5927) = 10	i(6353) = 10
i(5531) = 10	i(5939) = 10	i(6359) = 10
i(5557) = 10	i(5953) = 11	i(6361) = 10
i(5563) = 9	i(5981) = 10	i(6367) = 10
i(5569) = 9	i(5987) = 9	i(6373) = 10
i(5573) = 8	i(6007) = 10	i(6379) = 10
i(5581) = 10	i(6011) = 10	i(6389) = 12
i(5591) = 8	i(6029) = 10	i(6397) = 10
i(5623) = 12	i(6037) = 11	i(6421) = 10
i(5639) = 12	i(6043) = 10	i(6427) = 9
i(5641) = 12	i(6047) = 10	i(6449) = 9
i(5647) = 12	i(6053) = 10	i(6451) = 10
i(5651) = 12	i(6067) = 10	i(6469) = 8
i(5653) = 12	i(6073) = 10	i(6473) = 10
i(5657) = 13	i(6079) = 10	i(6481) = 10
i(5659) = 9	i(6089) = 10	i(6491) = 9
i(5669) = 12	i(6091) = 10	i(6521) = 10
i(5683) = 8	i(6101) = 10	i(6529) = 10
i(5689) = 10	i(6113) = 10	i(6547) = 8
i(5693) = 9	i(6121) = 10	i(6551) = 8
i(5701) = 10	i(6131) = 10	i(6553) = 10
i(5711) = 9	i(6133) = 10	i(6563) = 9
i(5717) = 8	i(6143) = 10	i(6569) = 9

Table 3 : Values of function $i(p)$: classes of prime numbers

i(6571) = 10	i(6983) = 12	i(7477) = 10
i(6577) = 9	i(6991) = 10	i(7481) = 12
i(6581) = 10	i(6997) = 10	i(7487) = 12
i(6599) = 8	i(7001) = 10	i(7489) = 10
i(6607) = 9	i(7013) = 10	i(7499) = 10
i(6619) = 9	i(7019) = 10	i(7507) = 10
i(6637) = 11	i(7027) = 10	i(7517) = 10
i(6653) = 10	i(7039) = 10	i(7523) = 10
i(6659) = 8	i(7043) = 11	i(7529) = 12
i(6661) = 9	i(7057) = 10	i(7537) = 10
i(6673) = 10	i(7069) = 10	i(7541) = 10
i(6679) = 10	i(7079) = 10	i(7547) = 10
i(6689) = 8	i(7103) = 10	i(7549) = 10
i(6691) = 10	i(7109) = 10	i(7559) = 10
i(6701) = 9	i(7121) = 10	i(7561) = 10
i(6703) = 10	i(7127) = 10	i(7573) = 9
i(6709) = 10	i(7129) = 10	i(7577) = 11
i(6719) = 11	i(7151) = 12	i(7583) = 9
i(6733) = 8	i(7159) = 10	i(7589) = 10
i(6737) = 8	i(7177) = 10	i(7591) = 12
i(6761) = 9	i(7187) = 10	i(7603) = 9
i(6763) = 10	i(7193) = 10	i(7607) = 9
i(6779) = 9	i(7207) = 10	i(7621) = 12
i(6781) = 12	i(7211) = 10	i(7639) = 10
i(6791) = 10	i(7213) = 10	i(7643) = 9
i(6793) = 8	i(7219) = 12	i(7649) = 10
i(6803) = 9	i(7229) = 10	i(7669) = 10
i(6823) = 10	i(7237) = 10	i(7673) = 12
i(6827) = 10	i(7243) = 10	i(7681) = 10
i(6829) = 8	i(7247) = 10	i(7687) = 9
i(6833) = 10	i(7253) = 12	i(7691) = 10
i(6841) = 10	i(7283) = 12	i(7699) = 10
i(6857) = 10	i(7297) = 10	i(7703) = 9
i(6863) = 9	i(7307) = 10	i(7717) = 8
i(6869) = 9	i(7309) = 10	i(7723) = 11
i(6871) = 8	i(7321) = 10	i(7727) = 12
i(6883) = 8	i(7331) = 10	i(7741) = 9
i(6899) = 9	i(7333) = 10	i(7753) = 8
i(6907) = 10	i(7349) = 12	i(7757) = 9
i(6911) = 8	i(7351) = 10	i(7759) = 10
i(6917) = 10	i(7369) = 10	i(7789) = 10
i(6947) = 12	i(7393) = 10	i(7793) = 10
i(6949) = 12	i(7411) = 10	i(7817) = 11
i(6959) = 12	i(7417) = 10	i(7823) = 11
i(6961) = 13	i(7433) = 10	i(7829) = 11
i(6967) = 12	i(7451) = 10	i(7841) = 10
i(6971) = 12	i(7457) = 10	i(7853) = 9
i(6977) = 12	i(7459) = 10	i(7867) = 10

Łukasiewicz Logics and Prime Numbers

i(7873) = 10   i(7879) = 10   i(7901) = 8
i(7877) = 12   i(7883) = 10   i(7907) = 10
                                            i(7919) = 10

Concluding remarks

Immanuel Kant, in his "Critique of Pure Reason" (1781), introduced the term "pure logic,"; the subtitle of G. Frege's celebrated work "Begriffsschrift ..." (1879) contains the words "pure (reines) thought"; E. Husserl, in his major work [Husserl (1913), 1982] goes to great lengths to bring out the essence of "pure consciousness"; lastly, D. Hilbert (with P. Bernays) in "Grundlagen der Mathematik" (1934) investigates "pure logic," meaning the first-order logic (it is well-known that the academic interests of Frege, Husserl, and Hilbert overlapped to a significant degree). The limitations of the first-order logic immediately become apparent, however, as soon as we try to use it to talk about different classes of structures – the first order logic can not distinguish, let alone define, the countable and the uncountable, it can not define the finite and the infinite. Therefore, ever more increasing attention has been paid over the last decades to the extensions of the first-order logic. The most prominent among these is probably the second-order logic, in which we can talk about virtually the whole of set theory, including the continuum hypothesis. A significant number of results concerning the extensions of the first-order logic was presented in a landmark volume "Model-Theoretic Logics" [Barwise and Feferman, 1985], where J. Barwise suggests that "There is no going back to the view that logic is first-order logic" (in this connection, see also [Sher, 1991]).

Moreover, it is still not clear how pure logic is related to the logic of human reasoning. W. Hodges in his article on elementary predicate logic revised for the second edition of "Handbook of Philosophical Logic" ([Hodges, 2001, ch. 28]) suggests that the connection, if it exists at all, is rather tenuous. (Let's note that the problem of the interconnection of pure consciousness and human consciousness is the one that occupied Husserl for nearly the whole of his life.) Then, it is interesting to understand if there is any link between human thought and the extensions of pure logic, as well as of its "core."

By the "core" of pure logic we mean the classical propositional logic C_2. What happens when this "core" is extended is what this book is partly about. The results by D'Ottaviano and Epstein, as well as by Anshakov and Rychkov, mentioned in the book indicate that Łukasiewicz three-valued, as well as any Łukasiewicz finite-valued logic, are not restrictions (as they are usually viewed), but in a sense, extensions of C_2. The repercussions of such an extension are quite serious as well as rather surprising. The extension of the very basic logical universe resulted in the logics of continual nature; in the possibility to characterize, structure, and describe classes of prime numbers. Are all of these required for the logical reasoning? On the other hand, the problem of fatalism and free will is also of continual nature, which usually goes unnoticed. As we wrote in this book for the refutation of the doctrine of logical fatalism, Łukasiewicz, without being aware of it, abandoned discreteness for continuity. But what precisely is the connection with prime numbers? Is the mystery of the logical universe somehow connected to the mystery of the distribution of prime numbers? If it is, Łukasiewicz logics must be a link between them.

References

[Adamatzky, 2004] A. Adamatzky. Review of [Karpenko, 2000]. *The Journal of Multiple-Valued Logic and Soft Computing*, 10 (3): 309-314, 2004.

[Anshakov and Rychkov, 1994] O. Anshakov and S. Rychkov. On finite-valued propositional logical calculi. *Notre Dame Journal of Formal Logic*, 36, No. 4: 606-629, 1994.

[Ayoub, 1963] R. Ayoub. *Introduction to the Analytic Theory of Numbers*. Providence, 1963.

[Barton, 1979] S. Barton. The functional completeness of Post's m-valued propositional calculus. *Zeitschrift für mathematische Logic und Grundlagen der Mathematik*, 25, H. 5: 445-446, 1979.

[Barwise and Feferman (eds.), 1985] J. Barwise and S. Feferman, editors. *Model-Theoretic Logics*. Springer-Verlag, Berlin, 1985.

[Beavers, 1993] G. Beavers. Extensions of the $\aleph_0$-valued Łukasiewicz propositional logic. *Notre Dame Journal of Formal Logic*, 34, No. 2: 251-262, 1993.

[Becchi, 2002] A. Becchi. Logic and determinism in Jan Łukasiewicz's philosophy. (http://www.philos.unifi.it/documenti/logic_determinism_lukasiewicz.pdf).

[Bennet, 1974] J. Bennet. Review of Taylor [1962]. *The Journal of Symbolic Logic*, 39, No. 2: 362-364, 1974.

[Blok and Pigozzi, 1989] W. J. Blok and D. Pigozzi. *Algebraizable Logics* (monograph). Memoirs of the American Mathematical Society. No. 396, 1999.

[Bochvar, 1981] D. A. Bochvar. On a three-valued calculus and its application to analysis of paradoxes of classical extended functional calculus. *History and Philosophy of Logic*, 2, 1981.

[Bochvar and Finn, 1972] D. A. Bochvar and V. K. Finn. On many-valued logics that permit the formalization of analysis of antinomies, I. In D.A. Bochvar, ed. *Researches on Mathematical Linguistics, Mathematical Logic and Information Languages*, pp. 238-295. NAUKA Publishers, Moscow, 1972 (in Russian).

[Bochvar and Finn, 1976] D. A. Bochvar and V. K. Finn. Some additions to papers on many-valued logics. In D.A. Bochvar and W.N. Grishin, editors. *Researches on Set Ttheory and Non-Classical Logics*, pp. 238-295. NAUKA Publishers, Moscow, 1976 (in Russian).

[Bolker, 1970] E. D. Bolker. *Elementary Number Theory. An Algebraic Approach*. W.A. Benjamin, New York, 1970.

[Burton, 1976] D. M. Burton. *Elementary Number Theory*. Allyn and Bacon, Boston etc., 1976.

[Byrd, 1979] M. Byrd. A formal interpretation of Łukasiewicz's logics. *Notre Dame Journal of Formal Logic*, 20, No. 2: 366-368, 1979.

[Caleiro, Carnielli, Coniglio and Marcos, 2005] C. Caleiro, W. Carnielli, M. Coniglio and J. Marcos. Two's company: "The Hamburg of many logical values. In J.-Y. Béziau, editor. *Logica Universalis*, pp. 169-189. Birkhäuser, Basel, 2005.

[Carmichael, 1922] R. D. Carmichael. Note on Euler's φ-function. *Bull. Amer. Math. Soc.*, 28: 109-110, 1922.

[Chang, 1958] C. C. Chang. Algebraic analysis of many-valued logics. *Transactions of the American Mathematical Society*, 88: 467-490, 1958.

[Chang, 1959] C. C. Chang. A new proof of the completeness of the Łukasiewicz axioms. *Transactions of the American Mathematical Society*, 93: 74-80, 1959.

[Church, 1956] A. Church. *Introduction to Mathematical Logic*. Vol. 1. Princeton Univ. Pres, Princeton, 1956.

[Cignoli, 1982] R. Cignoli. Proper *n*-valued Łukasiewicz algebras as *S*-algebras of Łukasiewicz *n*-valued proposional calculi. *Studia Logica*, 41, No. 1: 3-16, 1982.

[Cignoli, D'Ottaviano and Mundici, 2000] R. L. O. Cignoli, I. M. L. D'Ottaviano and D. Mundici. *Algebraic Foundation of Many-Valued Reasoning*. Trends in Logic. Vol. 7. Kluwer, Dordrecht, 2000.

[Copi and Cohen, 1998] I. M. Copi and C. Cohen. *Introduction to Logic*. Prentice Hall, New Jersey, 1998 (10th ed.)

[da Costa, Béziau and Bueno, 1996] N. C.A. da Costa, J-Y. Béziau and O. A. S. Bueno. Malinowski and Suszko on many-valued logics: On the reduction of many-valuedness. *Modern Logic*, 6, No. 3: 272-299, 1996.

[Epstein, 1990]. R. L. Epstein. *The Semantic Foundations of Logic*. Vol. 1: Propositional logic. Kluwer, Dordrecht, 1990 (2nd ed. in 1995).

[Erdös, 1958] P. Erdös. Some remarks on Euler's φ-function. *Acta Mathematika*, 4: 10-19, 1958

[Evans and Schwartz, 1958] T. Evans and P. B. Schwartz. On Slupecki T-functions. *The Journal of Symbolic Logic*, 23: 267-270. 1958.

[Finn, 1969] V. K. Finn. The precompleteness of a class of functions that correspond to the three-valued logic of J. Łukasiewicz. *Naucno-Tehnicheskaja Informacia* (VINITI), Ser.2. No. 10: 35-38, 1969 (in Russian).

[Finn, 1970] V. K. Finn. On classes of functions that corresponded to the n-valued logics of J. Łukasiewicz. *Proceedings of the Educational Institutions of the*

Central Area of the RSFSR. Section of Algebra, Mathematical Logic and Computer Science. Ivanovo, 1970 (in Russian).

[Finn, 1974] V. K. Finn. An axiomatization of some three-valued logics and their algebras. In *Philosophy and Logic*, pp. 398-438. NAUKA Publishers, Moscow, 1974 (in Russian).

[Finn, 1975] V. K. Finn. Some remarks on non-Postian logics. In *Vth International Congress for Logic, Methodology and Philosophy of Science. Contributed Papers.* Section 1, pp. 9-10. Ontario, Canada, 1975.

[Font, Jansana and Pigozzi, 2003] J. M. Font, R. Jansana and D. Pigozzi. A survey of abstract algebraic logic. *Studia Logica*, 74, No. 1/2: 13-97, 2003.

[Ford, 1998] K. Ford. The distribution of totients. *Ramanujan Journal*, 2: 67-151, 1998.

[Ford, 1999] K. Ford. The number of solutions of phi(x) = m. *Annals of Mathematics*, 1, 150: 283-311, 1999.

[Glaisher, 1940] J. W. L. Glaisher. *Mathematical Tables.* Vol. VIII. Number-Divisor Tables II, pp. 64-71. Cambridge, 1940.

[Göbel, 1980] F. Göbel. On 1-1 correspondence between rooted trees and natural numbers. *Journal of Combinatorial Theory.* Series B, 29, No 1: 141-143, 1980.

[Grigolia, 1977] R. Grigolia. Algebraic analysis of Łukasiewicz-Tarski's n-valued logical systems. In [*Wójcicki and Malinowski* (eds.), 1977, pp. 81-92].

[Gupta, 1981] H. Gupta. Euler's totient function and its inverse. *Indian Journal of Pure and Applied Mathematics*, 12: 22-30, 1981.

[Guy, 1994] R. K. Guy. *Unsolved Problems in Number Theory*, 2nd ed. § C1. Spring-Verlag, New York, 1994.

[Hales, 1971] A. W. Hales. Combinatorial represantations of Abelian groups. *Proc. of Symposia in Pure Mathematics*, 19: 105-108, 1971.

[Harary, 1994] F. Harary. *Graph Theory*, pp. 187-190 and 232. Addison-Wesley, Reading, MA, 1994.

[Hardy, 1999] G. H. Hardy. *Ramanujan: Twelwe Lectures on Subjects Suggested by His Life and work*, 3rd ed. Chelsea, New York, 1999.

[Hendry, 1980] H. E. Hendry. Functional completeness and non-Łukasiewiczian truth functions. *Notre Dame Journal of Formal Logic*, 21, No 3: 536-538, 1980.

[Hendry, 1983] H. E. Hendry. Minimally incomplete sets of Łukasiewiczian truth functions. *Notre Dame Journal of Formal Logic*, 24, No. 1: 146-150, 1983.

[Hendry and Massey] H. E. Hendry and G. J. Massey. On the concepts of Sheffer functions. In *The logical way of doing things*, pp. 179-293. New Haven and London, 1969.

[Herzberger, 1977] H. G. Herzberger. Tertium without plenum. In *International Symposium on Multiple-Valued Logic*. 7th, pp. 92-94. Charlotte, 1977.

[Hilbert und Bernays, 1968] D. Hilbert und P. Bernays. *Grundlagen der Mathematik*. I. Springer-Verlag, Berlin, 1968.

[Hodges, 2001] W. Hodges. Elementary predicate logic. In D. Gabbay and F. Guenthner, editors. *Handbook of Philosophical Logic*, pp. 1-127. Second Edition. Vol. 1. Kluwer,Dordrecht, 2001.

[Hulanicki and Swierckowski, 1960] A. Hulanicki and S. Swierckowski. Number of algebras with a given set of elements. *Bull. Acad. Polon., Sci. Sér. Sci. Math. Astronom. Phys*, 8:283-284, 1960.

[Husserl, 1982] E. Husserl. *Ideas Pertaining to a Pure Phenomenology and to a Phenomenological Philosophy - First Book: General Introduction to a Pure Phenomenology*. The Hague, Nijhoff, 1982.

[Jablonski, 1958] S. V. Jablonski. Functional constructions in k-valued logics. In *Studies of V.A. Steklov Mathematical Institute*, 51: 5-142. Moscow, 1958 (in Russian).

[Janov and Muchnik, 1958] Yu. I. Janov and A. A. Muchnik. On the existence of k-valued closed classes that have no bases. *Dokl. Akad. Nauk SSSR*, 127: 44-46, 1958 (in Russian).

[Janowskaja, 1959] S. A. Janowskaja. Mathematical logic, and foundations of mathematics. In *Mathematics in the USSR during 40 Years*, ch. 13. NAUKA Publishers, Moscow, 1959 (in Russian).

[Jaśkowski (1936). 1967] S. Jaśkowski. Investigations into the system of intuitionistic logic. In: S. McCall (ed.) *Polish logic: 1920–1939*, pp. 259-263. Clarendon Press, Oxford, 1967.

[Jordan, 1963] Z. A. Jordan. Logical determinism. *Notre Dame Journal of Formal Logic*, 4: 1-38, 1963.

[Karpenko, 1982a] A. S. Karpenko. Aristotle, Łukasiewicz and factor-semantics. *Acta Philosophica Fennica*, 35: 7-21, 1982.

[Karpenko, 1982b] A. S. Karpenko. Characteristic matrix for prime numbers. In *The Sixth All-Union Conference on Mathematical Logic. Abstract*, p. 76. Tbilisi, 1982 (In Russian).

[Karpenko, 1983a] A. S. Karpenko. Factor-semantics for n-valued logics. *Studia Logica*, 42, No. 2/3: 179-185, 1983.

[Karpenko, 1983b] A. S. Karpenko. The sequence of precomplete Łukasiewicz's logics and graphs for prime numbers. In *Proceedings of the Reseach Logical Seminar of Institute of Philosophy Russian Academy of Sciences*, pp. 103-111. Moscow, 1983 (in Russian).

[Karpenko, 1986] A. S. Karpenko. A hypothesis on the finitness of graphs for Łukasiewicz's precomplete logics (graphs for prime numbers). *Bulletin of the Section of Logic*, 15, No. 3: 102-108, 1986.

[Karpenko, 1988] A. S. Karpenko. *T-F*-sequences and their sets as truth-values. In M. Istvan et al., eds. *Intensional Logic, History of Philosophy, and Methodology*, pp. 109-119. Budapest, 1988.

[Karpenko, 1989a] A.S. Karpenko. Characterization of prime numbers in Łukasiewicz's logical matrix. *Studia Logica*, 48, No. 4: 465-478, 1989.

[Karpenko, 1989b] A. S. Karpenko. Finite-valued Łukasiewicz's logics: Algebra-logical properties of prime numbers. In V.A. Smirnov (ed.). *Sintactic and Semantics Investigation of Non-Extentional Logics,* pp. 170-205. NAUKA Publishers, Moscow, 1989 (in Russian).

[Karpenko, 1990] A. S. Karpenko. *Fatalism and Contingency of Futures. Logical Analysis.* NAUKA Publishers, Moscow, 1990 (in Russian).

[Karpenko, 1994] A. S. Karpenko. Sheffer's stroke for prime numbers. *Bulletin of the Section of Logic*, 23, No. 3: 126-129, 1994.

[Karpenko, 1995] A. S. Karpenko. Sheffer stroke for prime numbers. *Logical Investigation*, 3: 292-313. Moscow, 1995 (in Russian).

[Karpenko, 1996] A. S. Karpenko. The class of precomplete Łukasiewicz's many-valued logics and the law of prime number generation. *Bulletin of the Section of Logic*, 25. No 1: 52-57, 1996.

[Karpenko, 1997a] A. S. Karpenko. The law of prime numbers generation: Logic and computer realization. *International Conference on Information and Control. Proceedings*. Vol. 2, pp. 494-495. St. Peterburg, 1997.

[Karpenko, 1997b] A. S. Karpenko. *Many-Valued Logics* (monography). In "Logic and Computer". Vol. 4. NAUKA Publishers, Moscow, 1997 (in Russian).

[Karpenko, 1999] A. S. Karpenko. Characterization of classes natural numbers by logical matrices. In *Proceedings of the Reseach Logical Seminar of Institute of Philosophy Russian Academy of Sciences*, pp. 217-225. Moscow, 1999 (in Russian).

[Karpenko, 2000] A. S. Karpenko. *Łukasiewicz's Logics and Prime Numbers.* NAUKA Publishers, Moscow, 2000 (in Russian).

[Kleene, 1952] S.C. Kleene. *Introduction to Metamathematics*. D. Van Nostrand Company, New York, 1952.

[Lal and Gillard, 1968] M. Lal and P. Gillard. *Table of Euler's phi function, $n \le 10^5$*. Neufoundland. (Deposited in the UMT file), 200 pp., paperbound, 1968.

[Łukasiewicz, 1963] J. Łukasiewicz. *Elements of Mathematical Logic*. N.Y., 1963.

[Łukasiewicz, 1970] J. Łukasiewicz. *Selected works*. North-Holland, Amsterdam, 1970.

[Łukasiewicz, 1970a] J. Łukasiewicz. Farewell lecture by professor Jan Łukasiewicz, delivered in the Warsaw University Lecture Hall on March 7, 1918. In: [*Łukasiewicz*, 1970, pp. 84-86].

[Łukasiewicz, 1970b] J. Łukasiewicz. On three-valued logic. In [*Łukasiewicz*, 1970, pp. 87-88].

[Łukasiewicz, 1970c] J. Łukasiewicz. On determinism. In [*Łukasiewicz*, 1970, pp. 110-128].

[Łukasiewicz, 1970d] J. Łukasiewicz. A numerical interpretation of theory propositions. In [*Łukasiewicz*, 1970, pp. 129-130].

[Łukasiewicz, 1970e] J. Łukasiewicz. Philosophical remarks on many-valued systems of propositional logic. In: [*Łukasiewicz*, 1970, pp. 153-178].

[Łukasiewicz and Tarski, 1970] J. Łukasiewicz and A. Tarski. Investigations into the sentential calculus. In [*Łukasiewicz*, 1970, pp. 131-152].

[Lyndon, 1951] R. Lyndon. Identities in two-valued calculi. *Trans. Amer. Math. Soc.*, 71: 457-465, 1951.

[Maier and Pomerance, 1988] H. Maier and C. Pomerance. On the number of distinct values of Euler's φ-function. *Acta Arithmetica*, 49: 263-275, 1988.

[Malinowski, 1977] G. Malinowski. Classical characterization of n-valued Łukasiewicz calculi. *Reports on Mathematical Logic*, 9: 41-45, 1977.

[Malinowski, 1993] G. Malinowski. *Many-valued logics*. Clarendon Press, Oxford, 1993.

[Matiyasevicz, 1993] Y. V. Matiyasevicz. *Hilbert's Tenth Problem*. MIT Press, Cambridge, MA, 1993.

[McCall, 1968] S. McCall. Review of S.M. Cahn 'Fate, Logic and Time'. *Journal of Philosophy*, 65: 742-746, 1968.

[McKinsey, 1936] J. C. C. McKinsey. On the generation of the functions Cpq and Np of Łukasiewicz and Tarski by means of the single binary operation. *Bulletin of the American Mathematical Society*, 42: 849-851, 1936.

[McNaughton, 1951] R. McNaughton. A theorem about infinite valued sentential logic. *The Journal of Symbolic Logic*, 16: 1-13, 1951.

[Minari, ?] P. Minary. A note on Łukasiewicz's three-valued logic. (http://www.philos.unifi.it/materiali/preprint/wajsberg.pdf).

[Monteiro, 1967] A. Monteiro. Construction des algèbres de Łukasiewicz trivalentes dans les algèbres de Boole monadiques, I. *Mathematica Japonica*, 12: 1-23, 1967.

[Mundici, 1994] D. Mundici. A constructive proof of McNaughton's theorem in infinite-valued logic. *The Journal of Symbolic Logic*, 59, No. 2: 596-602, 1994.

[Ono and Komori, 1985] H. Ono and Y. Komori. Logics without the contraction rule. *The Journal of Symbolic Logic*, 50: 169-201, 1985.

[Pomerance, 1980] C. Pomerance. Popular values of Euler's φ-function. *Mathematika*, 27. No. 1: 84-89, 1980.

[Post, 1921] E. L. Post. Introduction to a general theory of elementary propositions. *American Journal of Mathematics*, 43, No. 3: 163-185. (Reprinted in: van J. Heijenoort, ed. *From Frege to Gödel: A Source Book in Mathematical Logic*, 1879-1931, pp. 264-283. Harvard Univ. Press, Cambridge (Mass.), 1967).

[Post, 1941] E. L. Post. *Two-Valued Iterative Systems*. Annals of Mathematical Studies. Princeton Univ. Press. Vol. 5, 1941.

[Prijatelj 1996] A. Prijatelj. Bounded contraction and Gentzen-style formulation of Łukasiewicz logic. *Studia Logica*, 57: 437-456, 1996.

[Ratsa, 1982] M. F. Ratsa. On functional completeness in intuitionistic logic of propositions. *The Proberms of Cybernetics*, 39: 107-150, 1982 (in Russian).

[Rasiowa, 1974] H. Rasiowa. *An Algebraic Approach to Non-classical Logics*. PWN, Warszawa, 1974.

[Rechman F. 1977] F. Rechman. On Euler's φ-function. *Islamabad Journal of Science*, 4. No. 1-2, 1977 (MR 80g: 10005).

[Rescher N. 1969] N. Rescher. *Many-Valued Logic*. McGraw-Hill Book Company, New York, 1969.

[Ribenboim, 1996] P. Ribenboim. *The New Book of Prime Number Records*. Springer-Verlag, New York, 1996.

[Ribenboim, 1997] P. Ribenboim Are there functions that generate prime numbers? // *The College Mathematics Journal*, 25, No. 5: 352-359. 1997.

[Richstein, 2000]. J. Richstein. Goldbach conjecture up to 4×10^{14}. *Mathematics of Computation*, 70, No. 236: 1745-1749, 2000.

[Rine, 1974] D. C. Rine. Review of [Urquhart, 1973]. *Mathematical Review*, 47, No. 2: 274-275, 1974.

[Rose, 1968] A. Rose. Binary generators for the *m*-valued and $\aleph_0$-valued Łukasiewicz propositional calculi. *Composito Mathematica*, 20: 153-169, 1968.

[Rose, 1969] A. Rose. Some many-valued propositional calculi without single generators. *Zeitschrift für mathematische Logic und Grundlagen der Mathematik*, 15: 105-106, 1969.

[Rosenberg, 1970] I. Rosenberg. Über die funktionale Vollständigkeit in der mehrvertigen Logiken. *Rospravy Ceskoslovenske Academie ved. Rada matematickych a prirodnich ved. Praha*, 80, s. 4: 3-93, 1970.

[Rosenberg, 1978] I. Rosenberg. On generating large classes of Sheffer functions. *Aequationes Mathematicae*. Basel, 17: 164-181, 1978.

[Rosser and Turquette, 1952] J. B. Rosser and A. R. Turquette. *Many-valued Logics*. North-Holland, Amsterdam, 1952 (2nd ed. 1958).

[Routley and Meyer, 1976] R. Routley and R. Meyer. Every sentential logic has a two valued semantics. *Logique et Analyse*, 74-75-76: 355-356, 1976.

[Ruskey, 1995-2003] F. Ruskey. *Information on rooted trees*. htpp://www.theory.csc.uvic.ca/~cos/inf/tree/RootedTree.html.

[Rytin, 1999] M. Rytin. *Finding the inverse of Euler totient function*. htpp://mathsource.wolfram.com/Content/WhatsNew/0210-755.

[Schulz, 1982] P. Schulz. Enumerations of rooted trees with an application to group presentations // *Discrete Mathematics*, 41: 199-214, 1982.

[Scott, 1974] D. Scott. Completeness and axiomatizability in many-valued logics, in L. Hehkin *et al.*, editors. *Proceedings of the Tarski Symposium. Proceeding of Symposia in Pure Matematics*. Vol. 25: 411-436, 1974.

[Scott, 1976] D. Scott. Does many-valued logic have any use? In S. Körner (ed). *Philosophy of Logic*, pp. 64-74. Univ. of California Press, Berkeley and Los Angeles, 1976.

[Sher, 1991] G. Y. Sher. *The Bounds of Logic. A Generalized Viewpont*. The MIT Press, Cambridge, 1991.

[Sloane, 1999] N. J. A. Sloane. Sequences A039650. *An On-Line Version of the Encyclopedia of Integer Sequences*. http://www.research.att.com/~njas/sequences/eisonline.html.

[Sloane and Plouffe, 1995] N.J.A. Sloane and S. Plouffe. *The Encyclopedia of Integer Sequences*. Academic Press, San Diego, C.A., 1995.

[Słupecki, Bryll and Prucnal, 1967] J. Słupecki, J. Bryll, T. Prucnal. Some remarks on the three-valued logic of J. Łukasiewicz. *Studia Logica*, 21: 45-70, 1967.

References

[Smiley, 1976] T. Smiley. Comment on [Scott, 1976]. In [Scott, 1976], pp. 74-88, 1976.

[Spyroponlus, 1989] K. Spyroponlus. Euler's equation $\varphi(x) = k$ with no solution. *Journal of Number Theory*, 32, No. 2: 254-256, 1989.

[Suszko, 1975] R. Suszko. Remarks on Łukasiewicz's three-valued logics. *Bulletin of the Section of Logic*, 4, N0. 3: 87-90, 1975.

[Takagi, Nakashima and Mukajdono, 1999] N. Takagi, K. Nakashima and M. Mukajdono. Explicit Representation of many-valued Łukasievicz logic functions. *Multiple Valued Logic. An International Journal*, 4: 249-266, 1999.

[Tarski, 1930a] A. Tarski. Über einige fundamentale Begriffe der Metamathematik. *Comptes Rendus des Séances de la Société des Sciences et des Lettres de Varsovie*. Classe III, 23: 22-29 (see: On some fundamental concepts of metamathematics. In *[Tarski, 1956, pp. 30-37]*.

[Tarski, 1930b] A. Tarski. Fundamentale Begriffe der Methodologie der deduktiven Wissenschaften I. *Monatshefte für Mathematik und Physik*, 37: 361-404 (see: Fundamental concepts of the methodology of the deductive sciences. In *[Tarski, 1956, pp. 60-109]*.

[Tarski, 1956] A. Tarski. *Logic, Semantics, Metamathematics. Papers from 1923 to 1938*. Clarenden Press, Oxford, 1956 (2nd ed. Indianopolis, 1983).

[Taylor, 1962] R. Taylor. Fatalism. *The Philosophical Review*, 71: 56-66, 1962.

[Tokarz, 1974a] M. Tokarz. Invariant systems of Łukasiewicz calculi. *Zeitschrift für mathematische Logic und Grundlagen der Mathematik*, 20: 221-228, 1974.

[Tokarz, 1974b] M. Tokarz. A method of axiomatization of Łukasiewicz logics, *Studia Logica*, 33: 333-338, 1974. (*Reprinted in* [Wójcicki and Malinowski (eds.), 1977, pp. 113-118]).

[Tokarz, 1977] M. Tokarz. Degrees of completeness of Łukasiewicz logics. In *[Wójcicki and Malinowski (eds.), 1977, pp. 127-134]*.

[Tsuji, 1998] M. Tsuji. Many-valued logics and Suszko's thesis revisited. *Studia Logica*, 60, No. 2: 299-309, 1998.

[Tuziak, 1988] R. Tuziak. An axiomatization of the finite-valued Łukasiewicz calculus. *Studia Logica*, 47, No. 1: 49-55, 1988.

[Urquhart, 1973] A. Urquhart. An interpretation of many-valued logic. *Zeitschrift für mathematische Logic und Grundlagen der Mathematik*, 19: 111-114, 1973.

[Urquhart, 1986] A. Urquhart. Many-valued logic. In D.M. Gabbay and F. Guenthner, editors. *Handbook of Philosophical Logic*. Vol. III: Alternatives in Classical Logic, pp. 71-116. Reidel, Dordrecht, 1986. (Revised version reprinted as

"Basic many-valued logic" in D.M. Gabbay and F. Guenthner, editors. *Handbook of Philosophical Logic*, 2nd Edition, Vol. 2, pp. 249-295. Kluwer Academic Publishers, Dordrecht, 2001.).

[Wajsberg, 1977a] M. Wajsberg. Axiomatization of the three-valued calculus. In M. Wajsberg. *Logical Works*, pp. 12-29. Ossolineum, Wroclaw, 1977.

[Wajsberg, 1977b] M. Wajsberg. Contributions to meta-calculus of propositions I. In M. Wajsberg. *Logical Works*, pp. 89-106. Ossolineum, Wroclaw, 1977.

[Wang, 1984] Y. Wang. *Goldbach Conjecture*. World Scientific Publ., Singapore, 1984.

[Webb, 1936] D. L. Webb. Definition of Post's generalized negative and maximum in terms of one binary operation. *American Journal of Mathematics*, 42: 193-194, 1936.

[White, 1983] M. J. White. Time and determinism in the Hellenistic philosophical schools. *Archiv fürGeschichte der Philosopie*, 65: 40-62, 1983.

[Wilf, 1982] H. Wilf. What is answer? *American Mathematical Monthly*, 89: 289-292, 1982.

[Wolenski, 1989] J. Wolenski. *Logic and Philosophy in the Lvov-Warsaw School*. Kluwer Academic Publishers, Dordrecht, 1989.

[Wójcicki and Malinowski (eds.), 1977] R. Wójcicki and G. Malinowski. *Selected Papers on Łukasiewicz Sentential Calculi*. Ossolineum, Wrocław, 1977.

[Wójcicki, 1988] R. Wójcicki. *Theory of Logical Calculi: Basic Theory of Consequence Operations*. Reidel, Dordrecht, 1988.

[Woodruff, 1974] P. W. Woodruff. A modal interpretation of three-valued logic. *Journal of Philosophical Logic*, 3: 433-440, 1974.

[Zagier, 1977] D. Zagier. The first 50 million prime numbers // *The Mathematical Intelligencer*, 0: 7-19, 1977.

Author Index

Subject Index

CPSIA information can be obtained at www.ICGtesting.com

234784LV00001B/22/A